Plant Cell Biology

STRUCTURE AND FUNCTION

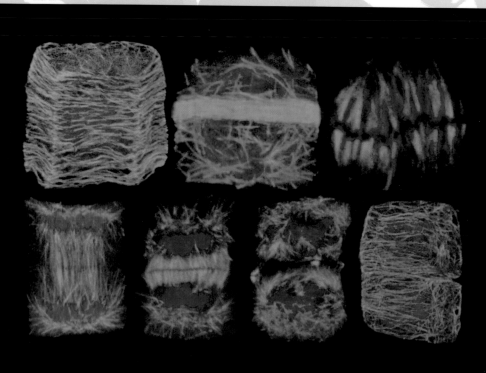

BRIAN E. S. GUNNING / MARTIN W. STEER

Plant Cell Biology

Structure and Function

Brian E. S. Gunning

Plant Cell Biology Group and Cooperative Research Centre for Plant Science
Research School of Biological Sciences
Australian National University
Canberra, Australia

Martin W. Steer

Department of Botany
University College Dublin
Dublin, Ireland

Jones and Bartlett Publishers

Sudbury, Massachusetts

Boston   London   Singapore

Editorial, Sales, and Customer Service Offices

Jones and Bartlett Publishers
40 Tall Pine Drive
Sudbury, MA 01776
508-443-5000
800-832-0034

Jones and Bartlett Publishers International
Barb House, Barb Mews
London W6 7PA
UK

This edition is published by arrangement with Gustav Fischer Verlag, Stuttgart. Original copyright © 1996 Gustav Fischer Verlag — ISBN 3-437-20534-X Title of German original: Gunning/Steer, *Bildatlas zur Biologie der Pflanzenzelle. 4. Auflage.*

Library of Congress Cataloging-in-Publication Data

Gunning, Brian E. S.
    [Bildatlas zur Biologie der Pflanzenzelle. English]
    Plant cell biology : structure and function / Brian
E. S. Gunning, Martin W. Steer.
        p.   cm.
    ISBN 0-86720-504-0 (paperback)   0-86720-509-1 (hardbound)
        1. Plant cells and tissues—Atlases.   2. Plant
ultrastructure—Atlases.   I. Steer, Martin W.   II. Title.
QK725.G8413 1995
581.87--dc20   95-37693
                                                CIP

Cover picture: Cells from wheat root tips, stained to show DNA (blue) and with anti-tubulin to show microtubules (green), imaged by confocal microscopy. From upper left: interphase cortical microtubules, mature preprophase band, metaphase spindle, early phragmoplast, late phragmoplast, cytokinesis almost complete, cortical arrays reinstated after division. See Plates 34 and 37 for further details.

Printed in the United States of America

00 99 98 97 96   10 9 8 7 6 5 4 3 2

Contents

Preface

The original edition of this book was an atlas of micrographs and legends called "Plant Cell Biology - an Ultrastructural Approach". It was published in 1975 and received consistent use in plant cell biology coursework, running to several printings in two languages over a long period. Constant requests have goaded us to undertake the present revision. We believe it to be timely for several reasons.

First, the relevance of structural aspects of plant cell biology has been greatly enhanced by the spectacular advances wrought by the "molecular revolution". Students of plant science now have new and powerful tools for exploring plant cells, melding structure with function in ways unheard of two decades ago. Molecular biologists are obtaining more and more exciting new information on structure-function relationships, in subject areas previously accessed almost solely by microscopists. Their micrographs of cells and tissues appear frequently on covers of the latest copies of molecular research journals. The massive new population of researchers working on plant cell and molecular biology brings a renewed need for a compact source of basic interpretations of plant cell structures and their biology, and examples of the methods by which they may be observed. We have sought to meet this need for both students and researchers, especially those whose work is extending from other areas of plant science into the realms of structural cell biology.

Second, microscopy itself has seen remarkable advances. Novel forms of light and electron microscopes are providing valuable new views and insights. As well as these new instruments there is a mushrooming armoury of techniques based on (e.g.) monoclonal antibodies, gene probes, reagents that report on physiological states within living cells, cryo-techniques, genetic transformants and mutants. These new tools have made enquiries into structure-function relationships a richer field of research than ever before. Perhaps the greatest advances lie in the addition of precise biochemical identifications and functional analyses to structural descriptions, and in the opportunities for studies of dynamic aspects of live, fully-functional cells.

In preparing the present revision, our major hurdle was that the original atlas was the pictorial component of a larger book: "Ultrastructure and the Biology of Plant Cells", in which more than 200 pages of text provided a complete background at that time. We have now aimed to produce a book that stands alone in the absence of accompanying text chapters. We have also aimed to update the contents to reflect 20 years of progress. Thus we have added 15 new plates to 45 of the original 49, and have augmented many of the originals. The micrographs and text diagrams now total more than 400, almost double the number in the original version. Cryo-microscopy, confocal microscopy, immuno-gold localisations, immuno-fluorescence microscopy, and *in situ* hybridisation are now featured. The text has been completely rewritten and greatly expanded. Our descriptions of the micrographs are given general introductions to set them in context. In turn they are used to introduce general concepts. A new title - "Plant Cell Biology: Structure and Function" - was necessary to reflect our broader textual coverage, and our inclusion of forms of microscopy over and above those traditionally known as "ultrastructural".

We are grateful for the encouragement that we have received from many colleagues; without them we would not have embarked on this venture. We are also grateful to colleagues who have donated micrographs to both this revision and the original edition, and for their enduring patience while they waited for their contributions to appear in print. They are acknowledged individually in appropriate figure legends.

Brian Gunning (Canberra)
Martin Steer (Dublin)

Introduction: Microscopy of Plant Cells

Knowledge of structure-function relationships is basic to our understanding of almost all biological phenomena. In this book we present 405 images and interpretations of plant cell structure, obtained by light and electron microscopy, and analyse the structural organisation that is revealed in them in terms of the biology of plants.

The Cell

Each cell is a community of subcellular components. Each type of component has its own particular set of functions. The individual parts could not survive for long outside the cell, but within the cellular environment they support each other so effectively that the cell *as a whole* is a viable entity. This subcellular cooperation not only ensures survival, but also provides for growth and multiplication of the cell (if given the necessary nutrients) and ultimately differentiation for a particular function.

Looking at a higher level of organisation, multicellular organisms are cooperative communities of cells, tissues and organs, all analogous to the subcellular components in that each contributes in a specialised way to the life of the system of which it is a part. No matter how complex the system, however, it is the cell that is the simplest, indivisible, unit which is viable - hence the common statement that *the cell is the unit of life*.

The quest to discover how cells work is one of the most exciting and important fields in biology. Cell biologists use many methods. Some take cells apart to study the components in isolation. Some go to even finer levels to look at biochemical properties of individual molecules, especially the enzymes that catalyse most life processes. However, all biologists realize that the cell is not just a soup of constituent parts and molecules. It is greater than the sum of its parts because it is an *organized* system. Hence all of the various methods for studying cells have a common focus at the level of structure. We need to know where the different kinds of molecule occur, how the sub-cellular components are constructed and how they function and interact with one another. It is fundamental to an understanding of biology to elucidate the structure-function relationships that generate the dynamic, sustainable entity that is the cell.

Observation of structure is therefore central to the study of cell biology. It provides a framework for understanding how cells take in and process raw materials, how they obtain and channel energy, how they synthesize the molecules they require for growth, how they multiply, how they develop specialized functions and how they interact with one another in tissues and organs. The pictures in this book focus on the biology of plant cells and tissues, as revealed by the structural organization that becomes visible when samples are magnified up to a few thousand times in the light microscope and up to a few hundred thousand times in the electron microscope.

Light Microscopy

In *conventional light microscopes* the patterns and colours produced by absorption of light by the specimen form magnified images, which reveal many basic structural features of cells. Plates 21a and 23a are examples of *light micrographs* which show general features of plant tissues. Additional information can be gained by using coloured stains to show up particular components with greater contrast (e.g. Plate 14a). Some of these are quite generic, e.g. stains for lipid, protein or carbohydrate. Others are exquisitely specific, e.g. probes made of nucleic acid that react only with particular genes or gene products (e.g. Plate 19c.d).

Other kinds of light microscope exploit variations in refractive index or optical path length in the specimen and can reveal structures that are nearly invisible in conventional instruments. The *phase contrast* microscope (see Plates 1 and 6a-f) and the *differential interference contrast* microscope (see Plates 38, 39) are useful for examining living cells because they can give informative images without requiring stains that may be toxic.

Fluorescence microscopes are especially valuable for relating biochemical activities to particular structures. Unlike forms of light microscope in which the light that illuminates the specimen goes on to form the image, the illuminating beam of fluorescence microscopes is used to excite fluorescent molecules in the specimen - either naturally occurring compounds, or dyes that attach to particular components. The excitation causes light to be emitted at a wavelength that is characteristic of the fluorescent substance. These emissions, or fluorescence, create an image from which the illuminating beam has been excluded by suitable filters. In addition, the illuminating beam is usually directed away from the microscope eyepieces (or camera) so that it does not interfere with the fluorescent image. Fluorescence microscopy has two main advantages over conventional light microscopy. First, because the fluorescent objects in the specimen are emitting light, they become visible even when the objects themselves are smaller than the usual resolution limit of light microscopes (Plates 34 and 35 give examples). The second advantage is that a vast array of fluorescent reagents is available, making it possible to reveal both structural features and physiological states, e.g. pH, electrical charge distributions, and concentrations of ions or other substances in living cells.

One specialized fluorescence microscope technique employs the great specificity of reactions between antibodies and their corresponding antigens. Antibodies can be raised by injecting substances (antigens) into rabbits, mice, or other animals with an immune system capable of making antibodies. The injected substances could, for example, be isolated and purified cellular components: e.g. an enzyme. Antibody molecules bind very specifically to their antigens, so they can be used as

1

highly specific reagents to locate the position of the original antigen in the cells or tissues from which it was extracted. The antibodies are made visible by attaching fluorescent labels to them. This is a very versatile procedure, capable of mapping the distribution of many substance in cells. Plates 34 and 35 give examples of *fluorescent antibody*, or *immuno-fluorescence*, localisations in plant cells.

One problem in light microscopy is that the in-focus image can be obscured by the out-of-focus images of objects above and below the focal plane. The *confocal laser scanning microscope* is a remarkable light microscope that reduces this problem. It excludes most out-of-focus light from the final image plane, giving beautifully clear images even from quite bulky specimens. These images are usually stored in digitized form in a computer memory. Successive *optical sections* from different levels of the specimen can be recombined to give three-dimensional reconstructions of the cell or tissue. This technique is especially useful in fluorescence microscopy, giving clear images of structures that have been labelled with fluorescent dyes (Plate 36q,r, 39b-k) or fluorescent antibodies (Plates 34,35,36a-p).

Electron Microscopy

The limit of resolution of the electron microscope is about 0.2nm (units of dimension are defined below), about 1000 times better than that of the light microscope (about 0.25μm). Although methods for preparing biological specimens do not always allow this level of detail to be seen, the three dimensional architecture of cells, their components, and even some large molecules, has been made visible to our eyes by electron microscopy.

The *transmission electron microscope* produces an image of the specimen by passing a beam of electrons through it. Electromagnetic fields manipulate and focus the beam, and the magnified image can be viewed directly on a fluorescent screen or recorded by black and white photography. Because electrons are easily deflected, or scattered, they are given a path that is as nearly as possible collision-free by evacuating most of the molecules of air from within the body of the instrument. It follows that specimens to be placed in the microscope must (a) be strong enough to stand up to the conditions of high vacuum, and (b) be thin enough to transmit sufficient electrons to give an image. These transmitted electrons form the light areas of the image on the fluorescent screen. Many of the electrons incident on the specimen are scattered or stopped, thereby creating dark areas in the final image, or *micrograph.*

In the *scanning electron microscope,* a finely focused beam of electrons is scanned in a regular (raster) pattern over the surface of the specimen. Electrons that are reflected, or caused to be emitted from the specimen, are collected and form an image on a conventional television cathode ray tube. Variations in the surface topography of the specimen lead to corresponding variations in the number of electrons collected as the beam sweeps over the specimen. These number variations are seen as brighter or darker patches on the TV screen. The final image portrays the three-dimensional surface in considerable detail. Examples are shown in Plates 5a; 28a,b; 48a,b; 52a,b,d,f, 56 and 57.

Specimen preparation

Several procedures have been used in the production of the electron micrographs presented here. Some very thin objects have been *shadow cast.* They were spread on a support film (a thin film of carbon or plastic, which is the electron microscopist's equivalent of the glass slide used by light microscopists) and sprayed with atoms of vapourised metal from a source placed to one side. All exposed surfaces accumulated a deposit that is dense to electrons, while sheltered areas remained un-coated and comparatively lucent to electrons. Just as objects imaged in aerial photographs can be measured and identified from the size and shape of their shadows, so it is with the shadow-cast electron microscope specimen (e.g Plates 4b and 11a, c, d). In some cases more detail is revealed if the electron-dense metal atoms are deposited from many directions by rotary shadowing (e.g. Plates 4c, 5e, and 6i)

Plates 4a, 5a, 9a, 16a, 22b,c and 33, exemplify an important variation on the shadow casting procedure. The specimen is frozen very rapidly so as to avoid distorting sub-cellular components, by the formation of ice crystals for example. Internal surfaces are then exposed by breaking open the frozen material. The fracture plane tends to follow lines of weakness, i.e. regions where there was not much water to start with, and which therefore have not produced strong ice. Cell membranes provide one such region, and so the exposed surface usually jumps from one expanse of membrane to another. The fracture usually passes along the *mid-plane* of cell membranes, fracturing them to reveal *internal* surfaces described as the PF (*protoplasmic face*, backing on to the cytoplasm) and EF (*extraplasmic face*, backing on to the non-cytoplasmic compartment, whether a lumen or outside the plasma membrane) (see diagram, Plate 33). Details of the surface topography of the fracture plane are then highlighted by subliming off some ice ("etching" the surface). Finally a replica of the surface is prepared in the form of a very thin layer of plastic and carbon. The replica faithfully follows the surface topography of the fracture plane, and the details can be viewed after it has been shadow cast. The whole operation is described as the *freeze-fracturing*, or *freeze-etching*, procedure. The method gives images that are especially trustworthy in being representations of material that was alive at the moment of freezing, and not altered since.

Most of the electron micrographs in this collection were obtained by the technique that is used more often than any other by cell biologists. The delicate cells are first chemically fixed, then encased in a hard material (embedded), and finally sectioned into slices thin enough to be stained and examined in the electron microscope.

Chemical fixation is a process in which the normal dynamic state of the cell components is interrupted by the application of *fixatives* - chemicals such as formaldehyde

2

or glutaraldehyde, which rapidly kill the cell and, by forming chemical bridges, cross-link the cellular molecules into a three-dimensional fabric rigid enough to stand up to the subsequent manipulations. The mode of action of a fixative can be demonstrated by adding it to a solution of a protein such as serum albumin: given suitable concentrations it is not many seconds before chemical cross-linking transforms the fluid solution into a solid mass. A second fixation step, which also functions in staining the specimen, is usually performed. The specimen is transferred from the first fixative to a solution of osmium tetroxide. This highly reactive substance reacts (to varying degrees) with many substances, and can be a cross-linking agent. Some cell components, lipids for example, take up more electron-dense osmium atoms than others and hence appear darker in the final image. In other words the osmium tetroxide is a differential stain as well as a fixative.

The fixed specimen is not usually strong enough to be sectioned, so next the water phase is replaced by a solvent and then by a liquid plastic. Epoxy resins are the most commonly used plastic embedding agents. After polymerisation in an oven to form a hard block, the specimen possesses the necessary strength for sectioning into extremely thin slices. "Thin" is a rather weak adjective in this context, and the term *ultra-thin* is usually used because the section thicknesses acceptable for conventional transmission electron microscopy are about one tenth of the wavelength of light (50nm). The sections have to be so thin that a piece of tissue 1mm thick could, in theory, be further sliced to yield 10,000-20,000 ultra-thin sections.

Although chemical fixation has been very widely used, it has severe disadvantages. Chief among these is that it does not bring rapid cellular processes to a halt sufficiently quickly to preserve subcellular details in their *in vivo* state. Accordingly, a technically more difficult, but better, method of preservation involving fast freezing is now preferred, and indeed is essential for especially delicate objects. For samples that are larger than a few μm in size, it is necessary to do this at high pressure to freeze the sample rapidly enough for it to be free of ice crystals. Once samples have been frozen they can be fractured for the freeze etch process (described above), or they can be infiltrated with resin and then sectioned (the *freeze substitution* process, e.g. Plates 46a,c; 57 and 58c,d), or they can even be sectioned while still frozen and then either freeze-dried before examination or examined in the frozen-hydrated state in an electron microscope that is equipped with a low temperature stage. Samples that have been fast frozen without chemical fixation have another advantage: their constituent molecules may show superior retention of their antigenicity, and this method of specimen preparation is therefore especially good for antibody localisation methods (see Plate 14 for an example).

As with light microscopy, there is a wealth of stains for use in electron microscopy. These stains must have large atomic nuclei to deflect electrons. In addition to osmium, mentioned previously, lead and uranium salts provide general staining, used in nearly all of the electron micrographs in this book. Other stains are based on compounds with silver, iron, and copper. The degree of specificity of staining extends all the way from general, non-specific stains to highly selective antibodies (labelled with tiny particles of gold, see Plates 14, 19 and 30) and nucleic acid probes.

Image interpretation

Electron micrographs of ultra-thin sections can be hard to interpret. A sound knowledge of the tissues at the light microscope level is an essential first step, but it is still hard to translate the two-dimensional image into the three-dimensional reality. It is rather like investigating a house, its rooms, its cupboards, and all their contents down to 1mm in size, by examining a 2cm thick slice of the whole building. Obviously it is desirable to look at many such slices, and they should, if at all possible, be cut in known planes or in sequences from which three dimensional reconstructions (e.g. Plate 18) can be made.

The dimensions of the world of ultrastructure are such that unfamiliar units, namely *micrometres* (symbol μm) and *nanometres* (symbol nm), are required:

1millimetre (mm) equals 1000μm

1μm equals 1000nm

or $1nm = 10^{-9}$metre; $1μm = 10^{-6}$ m; $1mm = 10^{-3}$metre

The true size of objects in a micrograph may be calculated using a simple and very useful rule of thumb. There are 1000μm in 1mm, therefore a 1μm object will appear to be 1mm in size when the magnification is x 1000. Scale-marker lines placed on a micrograph (e.g. Plate 1) enable the true dimensions of objects to be estimated at a glance. To place a scale-marker representing 1 μm on any micrograph, simply draw a line as many mm long as there are thousands in the magnification. Precise measurements are equally easily obtained, thus:

$$\frac{\text{size in micrograph (mm) x 1,000}}{\text{magnification}} = \text{true size (μm)}$$

or

$$\frac{\text{size in micrograph (mm) x 1,000,000}}{\text{magnification}} = \text{true size (nm)}$$

Cell structure and function

All cells possess certain basic biochemical systems that synthesize carbohydrates, proteins, nucleic acids, and many other types of molecule. All have an outer surface that provides protection by excluding harmful material in the external environment, while at the same time permitting the controlled import and export of other substances. This ensures stability of internal conditions. All cells have a store of information where the hereditary material embodies in a chemical form (DNA) instructions which guide the cells through the intricacies of their development and reproduction. All have devices which provide chemical energy that is utilized in the general maintenance of cellular integrity and in syntheses leading to growth and development. These attributes are fundamental to all living systems and the structural and functional similarities of plant and animal cells stem from them. There are, in addition, cellular features in which the two kingdoms differ, mostly deriving from two major events in the evolution of living organisms - the development of a cell wall and the acquisition of photosynthetic capabilities. The consequences for plant cell structure and function were far-reaching.

Plant cells vary in the extent to which different functions are developed, for, as with most multi-cellular organisms, plants exhibit division of labour. As a result of the varied requirements of maintaining life and supporting growth and development, specialized cells develop for protection, mechanical support, synthesis or storage of food reserves, transport, absorption and secretion, meristematic activity, reproduction, and the vital role of interconnecting the more specialized tissues.

Plants concentrate their processes of cell multiplication to permanently embryonic regions termed meristems, and the zones between meristems and the nearby mature tissues contain cells in intermediate stages of maturation. A comparison of a juvenile and a mature stage illustrates the great precision and specificity with which cell differentiation takes place behind a meristem. Plate 60 shows the central portion of an unusually "miniaturized" root at an early stage of development and Plate 51a depicts the same cell types, distributed in exactly the same geometrical pattern, but in a mature part of the same root. Six different types of cell have matured in their own characteristic fashions and at their own characteristic rates. All started from a population of comparatively uniform meristematic cells, differing from one another mainly in their positions in the meristem. In an organization of this sort there is clearly no such thing as a "typical plant cell", but meristematic cells must at least contain a basic set of components. They alone have not, or have only just, started to diversify by maturation, so it is logical to use them in an introductory survey of "the cell" (Plates 1-4, Fig.1), before examining the constituent parts in detail.

Plate 1 is a light micrograph, showing cells in a broad bean root tip that was fixed in glutaraldehyde, dehydrated, embedded in plastic, sectioned at about 1 μm thickness, and the section stained by a combination of procedures chosen to reveal as many as possible of the cell components. Finer detail is visible in Plates 2 and 3, which are electron micrographs of ultra-thin sections of cells in other root tips. Plate 4 gives a different view of the cell wall and the plasma membrane, obtained by freeze fracture and shadow casting methods that reveal surface textures. The final illustration in the introductory survey (Fig.1) attempts to overcome the artificial two-dimensional impression created by the micrographs. It is a stylized three-dimensional interpretation of that mythical entity, the "typical" plant cell. For the sake of clarity it is shown isolated from all the neighbours to which it should be joined. The components drawn within it are more symmetrical and simplified than they would be in life. Of necessity some have been enlarged in order to make them visible alongside their larger companions.

Plant tissues are composed of the non-living extra-cellular region and the living protoplasm of the cells proper. The former consists of intercellular spaces and cell walls. Each protoplast consists of a nucleus (or sometimes several nuclei) and the cytoplasm. Within these are the various membranous and non-membranous components which are described in the following glossary of names and outline descriptions. Numbers in brackets after each item indicate which of the first four plates give the best views of the structure in question; the letters in brackets refer to labels on Fig.1.

Cytoplasm: A collective term for everything outside the nucleus, out to and including the plasma membrane. Includes membranous and other inclusions, and also the general matrix, or *cytosol*, in which the cytoplasmic components reside. Excludes extracellular components (cell wall and other extracellular matrix material, intercellular spaces). (1-3).

Plasma membrane: The bounding membrane of the protoplast, normally in close contact with the inner face of the cell wall (2,3b,4a).

Nucleus: This is bounded by the nuclear envelope and contains genetic material in the form of chromatin, and the *nucleolus* (or, if more than one, *nucleoli*) in a matrix of *nucleoplasm*. (1,2).

Nucleoplasm: Everything enclosed by the nuclear envelope falls in the category of nucleoplasm, just as objects outside it are constitutents of the cytoplasm. The word is often, however, used to denote the ground substance in which the chromatin and nucleolus lie (1,2).

Chromatin: Contains the genetic material of the cell, i.e. information in the form of DNA that is passed from parent cell to daughter cell during the multiplication of cells and reproduction of the organism. It can exist in less dense (*euchromatin*) and more dense (*heterochromatin*) forms. During division of nuclei it is condensed into discrete units, *chromosomes*. (1).

Nucleolus: A mass of filaments and particles (NU), largely a sequence of identical repeating units of specialized genetic material together with precursors of ribosomes produced from that genetic information (1,2).

Nuclear envelope: A cisterna (a general term meaning a membrane-bound sac) wrapped around the contents of the nucleus (N). The space between the two membranous faces of the cisterna is the peri-nuclear space (1,2,3a).

Nuclear envelope pores: Elaborate perforations (NP) in the nuclear envelope, involved in transport between nucleus and cytoplasm and in the processing of messenger and ribosomal RNA molecules that are being exported from the nucleus (3a).

Endoplasmic reticulum: Membranous cisternae that ramify through the cytoplasm, occasionally connected to the outer membrane of the nuclear envelope. The bounding membrane segregates the contents of the cisterna from the cytoplasm. The outer face frequently bears attached ribosomes and polyribosomes (see below). Endoplasmic reticulum (ER) is described as *rough*, or granular (RER), and forms that lack ribosomes as *smooth*, or agranular (SER). A special form that lies just inside the plasma membrane is called *cortical* ER. ER cisternae may or may not have visible contents, which distend the cisternae when present in bulk (2, 3a, 3b).

Ribosomes: Small particles of RNA and protein lying free in the cytoplasm or else attached to the endoplasmic reticulum. They aggregate in clusters, chains, spirals, or other *polyribosome* configurations when they are engaged in protein synthesis (2, 3a).

Golgi bodies: The units of the Golgi apparatus of the cell. Each Golgi body (= Golgi stack) (G) consists of layered cisternae together with many small vesicles (VE, 1-5) that are involved in traffic to and from the Golgi apparatus and between its constituent cisternae (3a,b).

Vacuole: Compared with the surrounding cytoplasm, these are usually empty looking spaces (V), spherical when small. They are often very large, and can occupy 90% or more of the volume of the cell in mature tissues (1).

Tonoplast: The membrane that bounds a vacuole. Except for its position in the cell it looks very like the plasma membrane (3a, 3b).

Mitochondria: These pleiomorphic bodies (M) consist of a compartment, the *matrix,* surrounded by two membrane barriers, a *double envelope.* The outer membrane of the double envelope is more or less smooth, but the inner is thrown into many folds - *mitochondrial cristae* - that project into the matrix (2,3a).

Plastids: This is a group name for a whole family of cell components. In the young root-tip cells of the first three plates this group is represented by the structurally simplest member, which is called the *proplastid* (P). Proplastids are usually larger than mitochondria, but, like them, have a double membrane envelope surrounding (in these examples) a fairly dense ground substance, the *stroma.* Starch grains (ST) may be present in them (1, 3a). Other members of the plastid family, illustrated in later plates, are: *chloroplasts, etioplasts, amyloplasts,* and *chromoplasts.*

Microbodies: These (MB) are bounded by a single membrane, and are distinguished from vesicles by their size and dense contents (sometimes including a crystal) (2).

Microtubules: Except during cell division, these very narrow cylinders (MT) lie just inside the plasma membrane. The wall of the cylinder is made of protein and is *not* a cell membrane, though it may superficially resemble one in its thickness and density (2, 3a). Microtubules and microfilaments (below) are the major components of the *cytoskeleton* of plant cells.

Microfilaments: Fine fibrils, largely of filamentous *actin.* Plant actin is similar to its counterpart in animals, where it is one of the major constituents of muscle. Microfilaments are not illustrated in Plates 1-4.

Cell wall: This is a thin structure in meristematic cells, but it can be very massive and elaborate in mature cells. It is external to the living protoplast, but nevertheless contributes very significantly to the life of the plant cell; indeed, along with plastids, it is the major determinant of the lifestyle of plants. One of its main constituents is microfibrillar cellulose - the most abundant macromolecule on Earth (1,2,4).

Plasmodesmata: Narrow cytoplasmic channels (PD), bounded by the plasma membrane, which interconnect adjacent protoplasts through the intervening wall. The singular is plasmodesma (2,3b).

....

The micrographs in the 60 plates that follow have been selected to illustrate both structure and function at cellular and subcellular level, and dynamic aspects of plant cell biology. As well as providing descriptions, the legends introduce general concepts in the field.

Following the introductory survey (Plates 1-4), the remaining plates are in two main sections. The first section (Plates 5-44) covers subcellular components in the sequence: nucleus, endoplasmic reticulum, Golgi apparatus, vacuoles, mitochondria, plastids, the cytoskeleton and plasmodesmata. In the second section (Plates 45-60) the focus shifts from the biology of subcellular components to the biology of selected cell types, covering transfer cells, xylem, phloem, endodermis, epidermis, glands, male and female reproductive cells, leading up to the integration of cellular function in complex tissues.

Fig. 1

Diagram of a generalised plant cell cut open to show the three-dimensional structure of the principal components and their inter-relationships. For clarity they are not drawn to scale and some are illustrated by a few examples only (e.g. ribosomes). The letters used to label components are the same as those in parentheses in the list in the last two pages of the Introduction.

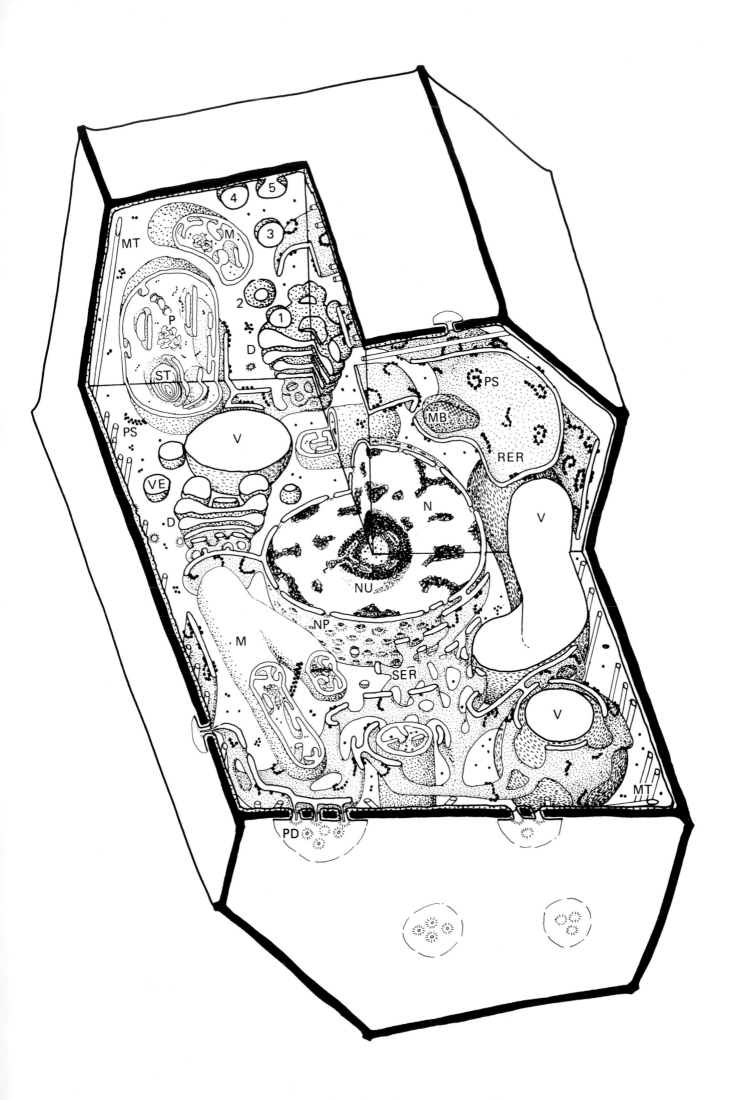

Plates 1-4 provide an introductory survey of the components of plant cells, starting with light microscopy (1), then using the higher resolution of the electron microscope to look at ultra-thin sections of a single cell (2) and parts of cells at higher magnification (3). Finally, other electron microscope techniques are used to give alternative views of the plasma membrane and cell wall at the cell surface (4).

1   The Plant Cell (1): Light Microscopy

The meristematic region of a broad bean (*Vicia faba*) root tip was chemically fixed using glutaraldehyde, embedded in glycol methacrylate plastic and sectioned with a glass knife at a section thickness of about 1μm. A section is viewed here by phase contrast microscopy at the best available resolution of the light microscope. The section was oxidised with periodic acid and reacted with the dye acriflavine. This procedure stains many carbohydrates yellow (e.g. in cell walls and starch grains). Subsequent immersion in iodine dissolved in potassium iodide gave a generally yellow-stained preparation. It was examined using the complementary wavelength, i.e. blue light, to obtain the best contrast and resolution. The magnification is x4,200, therefore 4.2mm in the micrograph represents a true dimension of 1μm. Since the thickness of the section was about 1μm, we are in effect looking through a slice 4.2mm in thickness, rather than at an infinitely thin 2-dimensional picture.

Each cell is outlined by its wall (CW). Thin regions of cell wall are sites where groups of intercellular connections, i.e. plasmodesmata, pierce the wall. There are no intercellular spaces in this particular group of cells. The major visible compartments of the cells are the numerous empty-looking vacuoles (V), which are small in meristematic cells, the cytoplasm, containing a variety of weakly and densely stained components, and the nuclei, which are numbered N-1 to N-4.

Each nucleus is separated from the cytoplasm by its nuclear envelope (NE). It is most clearly visible around nucleus N-1, which was at an early stage of mitosis when the tissue was fixed. In the other nuclei the speckles represent stained chromatin (CH). Before division the chromatin condenses to form discrete chromosomes (CHR in N-1), leaving the nuclear envelope relatively isolated and conspicuous prior to its breakdown at the onset of the next stage of mitosis. After cell division is complete the chromosomes uncoil again to regenerate the dispersed chromatin condition. A stage of this process is seen in nucleus N-2.

The large dense bodies in the nuclei are nucleoli (NL). The nucleolus in N-1 is lobed and irregular, but in the non-dividing nuclei its circular outline is indicative of a more-or-less spherical shape. Weakly-stained voids occur in most nucleoli. No nucleolus is seen in nucleus

N-3, but this does not mean that none is present. Sections are statistical samples of cells and tissues, and it is not to be expected that any one view of a cell will contain all of the possible subcellular components. For example, consider the dimensions of nucleus N-4, and assume that it and its nucleolus are spheres. Both are sectioned across their diameters, which at 4.2mm representing 1μm, are actually about 10μm and 4μm respectively. It would take 10-11 consecutive 1μm sections to pass from one face of the nucleus to the other, and only 4-5 of these would include portions of nucleolus. The remaining sections would not include any nucleolus (as in N-3).

The most clearly resolved cytoplasmic components are proplastids (PP), but these can only be identified with certainty if starch grains (e.g. arrow, top right) can be detected inside them by the combined staining with acriflavine and iodine-potassium iodide. Their varied profiles in the section (some elongated, some less so, some round particles) show that there is a population of randomly oriented proplastids in the cells. Some are more-or-less cylindrical and others may be more-or-less spherical. Only rarely does the full length of the elongated forms lie within the thickness of the section. Oblique sections of cylinders do not reveal the true length of the object.

Other cytoplasmic structures can be discerned but cannot be identified with certainty. The less densely stained particles doubtless include mitochondria (M?), and the very faint convoluted shadows (e.g. connecting the small arrows above N-1) are probably cisternae of endoplasmic reticulum, but this is a statement which can only be made with the benefit of hindsight in the light of electron microscope studies of these and similar cells. There is undoubtedly much more endoplasmic reticulum in these cells than can be detected by the method of specimen preparation that was used here. Cytoskeletal elements, Golgi bodies and microbodies must also be present but specific staining methods would be needed to identify them.

Plate 2 extends the amount of visible detail by taking advantage of the greater resolution of the electron microscope. An ultra-thin section of a meristematic cell is shown, stained by means of a general procedure that gives a overall view of subcellular components.

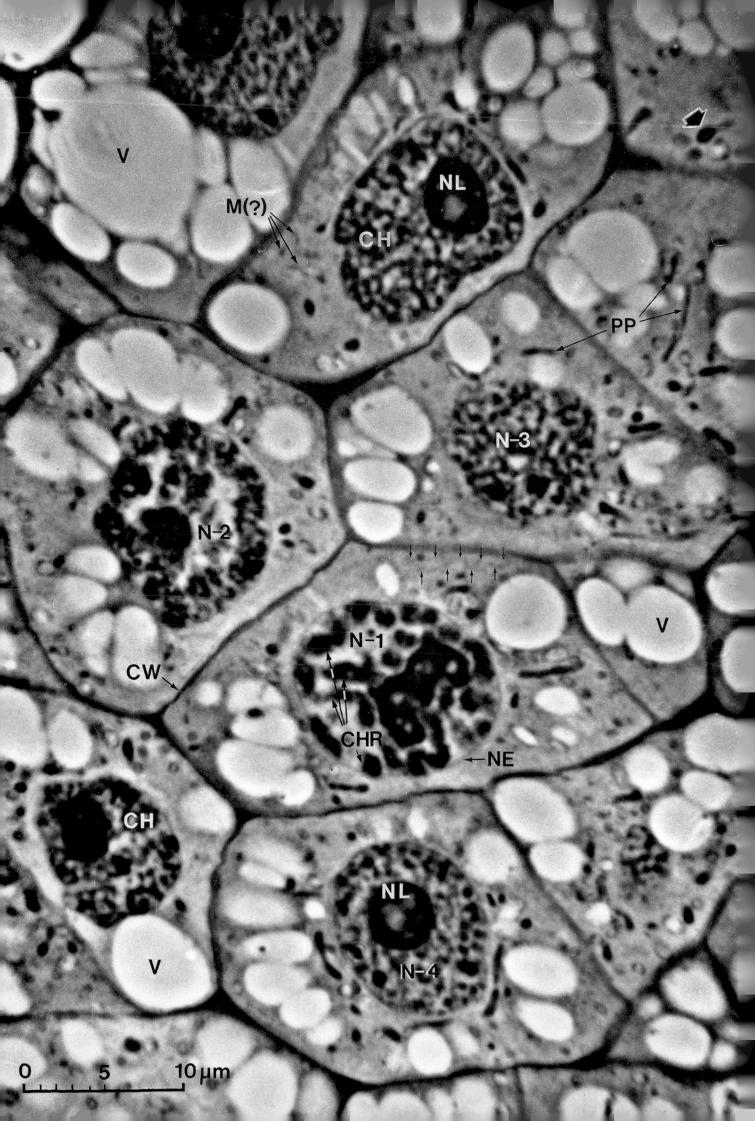

2   The Plant Cell (2): Overview by Electron Microscopy

This section sliced through the mid-region of a cell in the meristematic region of a root tip of cress *(Lepidium sativum),* and is viewed here by electron microscopy at a magnification of x20,000 after staining with uranium and lead salts to impart electron density. The section was about 75nm thick, so at this magnification the apparent thickness of the slice is about 1.5mm - relatively thin compared with Plate 1.

The higher resolution of the electron microscope brings to light many features not seen in the light micrograph of Plate 1. The plasma membrane (PM) is a densely stained profile, clearly distinguishable from the wall external to it (CW). The fibrillar texture of the cell wall is just visible at this low magnification. By and large the plasma membrane lies at right angles to the plane of the section, so that we are looking at it edge-on. It therefore appears as a dark line in the section. Its wrinkled outline is probably an artefact of chemical fixation, which is less effective at immobilising lipids than proteins. It is much smoother than this in preparations that have been preserved by rapid freezing (e.g. Plate 4).

The plasma membranes of adjacent cells pass through the intervening cell wall at plasmodesmata (PD). If we could see these channels of intercellular communication end-on, as in a section cut in the plane of the cell wall, they would appear as round profiles, extensions of the plasma membrane (see Fig. 1 in the Introduction and Plate 45). Because the section has a known thickness, we can estimate roughly how many plasmodesmata pierce the wall. Eleven plasmodesmata are wholly or partially included in the 250mm length of the wall at the left hand side. Since at a magnification of x20,000, 20mm represents $1\mu m$, this equates to 11 plasmodesmata in an area of wall equal to $12.5\mu m$ multiplied by the section thickness (75nm), i.e. $0.94\mu m^2$. This calculation shows that on average there is just over one plasmodesma per μm^2 of the surface of this type of cell. Note, however, that this is an over-estimate, because not all of the plasmodesmata in the count of 11 are complete, i.e. some are truncated by either the upper or the lower surface of the section.

The vacuoles (V) are unusually sparse and small in this view (cf. Plate 1) and their limiting membrane, the tonoplast, is not shown clearly. They lie in a densely particulate cytoplasm, each particle being a ribosome, about 20nm in diameter. Ribosomes either lie free in the cytoplasm or else are attached to the membranes of rough endoplasmic reticulum cisternae, giving them a characteristic beaded appearance (ER). In these cells most of the ER cisternae are narrow, and it should be realized that only those which lie at right angles to the plane of the section, or nearly so, are clearly visible.

Other components of the cytoplasm are present: mitochondria (M), proplastids (PP), Golgi bodies (G), and a microbody (MB). Note the difference in the staining density of mitochondria and proplastids. Although only a few of each of these components is visible in the section, they are undoubtedly numerous in the cell as a whole. The proplastids in this view do not contain starch grains.

A small region (outlined in the box at top left) is shown at magnification x40,000 in the inset (lower left) in order to make the cross sections of microtubules present there more obvious (arrows). In non-dividing cells, like this one, the microtubules are nearly all in the cell cortex, near the cytoplasmic face of the plasma membrane. The other major component of the cytoskeleton, microfibrils of actin, is not visible in this preparation.

The inner and outer membranes of the nuclear envelope (NE) are resolved. Nuclear envelope pores are only just visible at this magnification, so description of them is deferred to the next plate. The outer membrane of the nuclear envelope, like the endoplasmic reticulum, bears ribosomes.

The nuclear contents, perhaps most conspicuously different from the cytoplasm in the absence of membranes, consist of nucleolus (NL) and chromatin (CH) suspended in the general ground substance, or nucleoplasm. In this cell the chromatin is present mostly in a dispersed form (not labelled). There is also some dense heterochromatin (labelled CH). The ratio of heterochromatin to dispersed chromatin varies widely according to the nature of the cell and its stage of development. Here the high degree of dispersion probably reflects intensive transcription of genes, as would be expected in an actively growing and dividing cell. Heterochromatin is generally considered to be a less active form of the genetic material.

The nucleolus is an unusually large example, relative to the size of the nucleus, with fewer and smaller low density voids than those in Plate 1. It exhibits differentiation into regions that are predominantly granular (solid stars) and others in which substructure cannot be seen at this magnification, being composed of closely packed fine fibrils (open stars). The fibrils are RNA strands that will fold to become the RNA component of cytoplasmic ribosomes. The granular component is a mass of the same kinds of RNA strands, but now folded and complexed to various degrees with ribosomal proteins that have been imported into the nucleus from their site of manufacture in the cytoplasm. The granules will be exported from the nucleus through nuclear envelope pores to become cytoplasmic ribosomes. The open arrows point to strands of chromatin that ramify through the body of the nucleolus. These are the *nucleolar organiser* regions of the genetic material of the cell. They include many copies of the genes for ribosomal RNA.

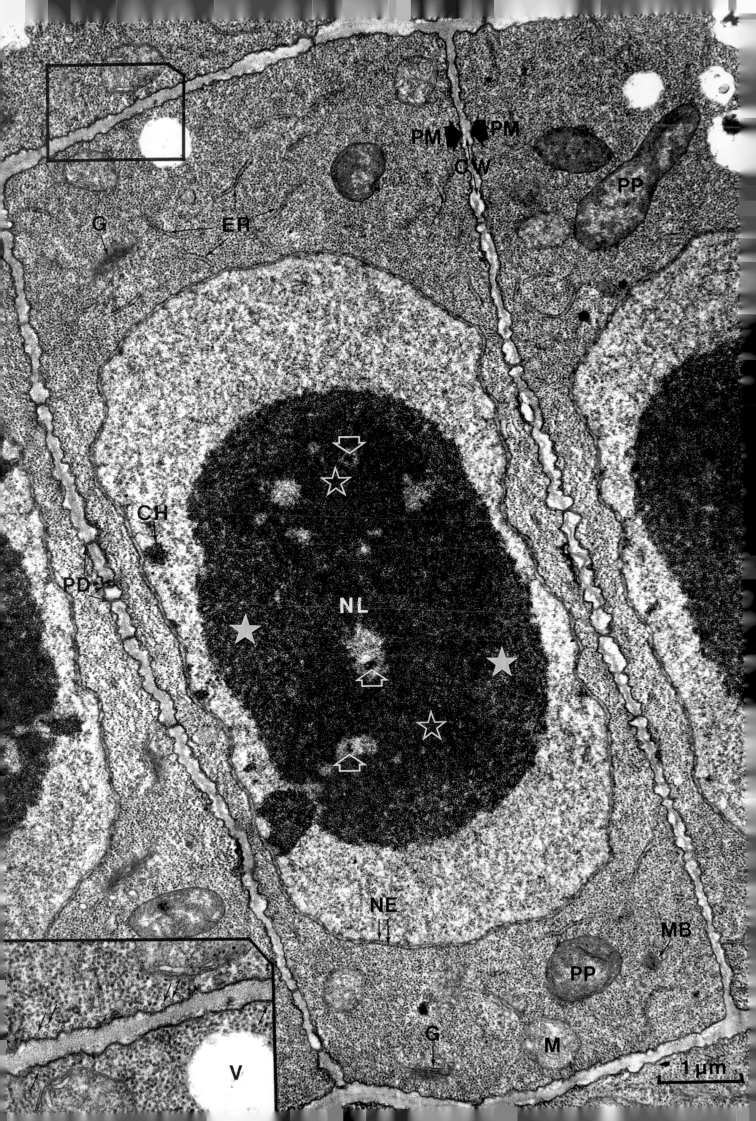

3 The Plant Cell (3): Ultrastructural Details

Plate 3a The same material as in Plate 2 is seen here at a higher magnification (x45,000). Parts of two cells are shown. The intervening cell wall (CW) is lined by the plasma membranes of the adjacent cells (PM) in a region without plasmodesmata.

The cisternae of endoplasmic reticulum (ER) are all of the rough category, with ribosomes studding their membranes. Ribosomes also lie free in vast numbers in the cytoplasm. One cisterna (ER*, lower right) is distended by the presence of accumulated contents, probably protein synthesised by the ribosomal machinery on the cytoplasmic face of the ER and injected through the membrane into the lumen of the cisterna.

The ER is not as sparse as might appear at first glance. It is merely that many of its cisternae happen to lie in or close to the plane of the section and therefore have a low contrast appearance. The arrowheads indicate the limits of one, barely discernible, obliquely-sectioned cisterna. Quantitative measurements show that the ER membrane system has a very large surface area, usually several μm^2 per μm^3 of cytoplasm.

The tonoplasts of the two vacuoles that enter the lower edge of the picture again illustrate the difference between the crisp profile of a membrane lying edge-on (T) and an indistinct, obliquely sectioned membrane (T*).

A small portion of a nucleus (N) is included at the right hand side of the micrograph. The two membranes and pores (open arrows) of the nuclear envelope (NE) are visible.

The successive cisternae of a Golgi stack (G) lie at right angles to the plane of the section, and many associated vesicles lie nearby in the cytoplasm, or else attached to the Golgi cisternae. Some of the vesicles are of the *coated* variety - their outer surface carries an investment of protein chains. Golgi stacks are not just piles of uniform flat cisternae. The successive cisternae can be seen to differ from one another in their structure (from left hand side to right hand side of the stack in this example). Although not visualised here, they also differ in their biochemical functions.

The double membrane envelope of proplastids (PP) and mitochondria (M) can be seen (squares). The inner membrane of the mitochondrial envelope is invaginated to give rise to numerous internal sacs (mitochondrial cristae), on whose surfaces lie enzyme and electron transport complexes that function in energy production and cell respiration. The internal membranes of the proplastids are less densely packed than those of the mitochondria. In root cells like this one the proplastids remain small and simple, but in leaf tissue they become very elaborate, with complex internal membrane systems that carry photosynthetic pigments.

Some microtubules (arrows) lie just internal to the plasma membrane, and two (circled) are unusually deep in the cytoplasm.

Plate 3b This high magnification picture (x100,000) illustrates aspects of the molecular organisation of cell membranes in a cell in the root tip of *Azolla*. The plasma membrane (PM) lines the cell wall (CW) and pierces it at plasmodesmata (PD). Other membranes included are the tonoplast (T), and cisternae of endoplasmic reticulum (ER) and a Golgi stack (G).

The material has an unusual appearance, but it has been selected for two reasons. One is that the cells contain some substance which spreads onto the surfaces of all membranes, probably during fixation. After staining it highlights a dark-light-dark appearance when the membranes lie edge-on in an ultra-thin section. The other reason is that these cells accumulate lipid in a form that (together with the staining effect) illustrates some of the attributes of cell membranes. A small droplet of concentrically-layered lipid (a *myelin figure* (MF)) is seen at the interface between cytoplasm and vacuole (V). Its layers are bimolecular sheets of lipid molecules lying back-to-back. The dark-light-dark stratification seen in individual cell membranes is good evidence that they too have this basic molecular architecture.

The insets show details enlarged from the same micrograph (x250,000). From left to right there are examples of myelin figure, plasma membrane, tonoplast, endoplasmic reticulum and a Golgi cisterna. Each circle encloses one short portion of triple-layered membrane. The stratification in individual membranes arises because the constituent lipid molecules are aligned in two opposed sheets. The hydrophobic tails of the lipids face each other in the interior of the membrane (lightly stained in the electron micrograph) and the hydrophilic head groups (which ususally attract more stain) are on the outer faces, in contact with the surrounding aqueous medium.

The stratified staining pattern thus reflects the general molecular organisation of cell membranes, based on bimolecular layers of lipid. However, most membranes also have protein constituents, though methods for staining ultra-thin sections do not usually reveal them. *Integral membrane proteins* consist of chains of hydrophobic amino acids largely embedded within the lipid environment of the membrane, usually with hydrophilic sites exposed at one or other (or both) of its faces. *Membrane-associated proteins* lie in the aqueous medium next to the membrane, anchored to it by hydrophobic amino acids that insert into the lipid environment of the membrane or by binding to integral membrane proteins. The two categories of membrane protein have many functions in cells, including trans-membrane transport systems, energy production, biosynthesis, reception of chemical and mechanical signals, and systems that transduce physical and chemical messages across the membrane. Methods for visualising or detecting the activity of some of these types of membrane protein are illustrated in other plates.

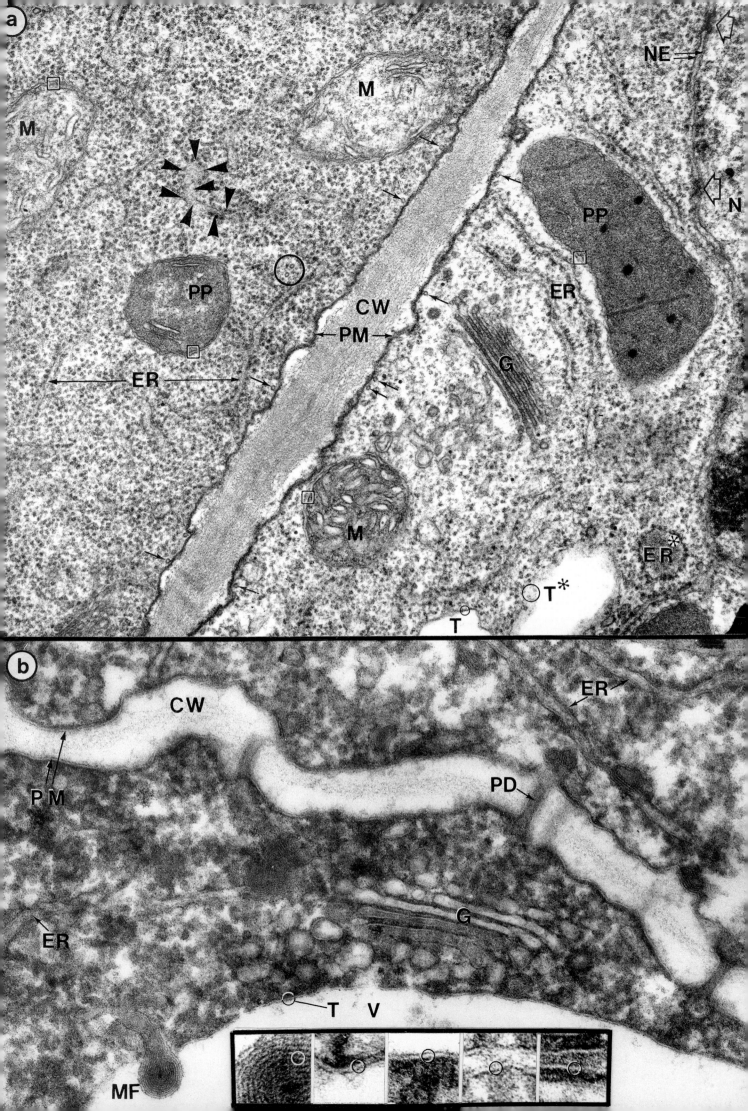

4   The Plant Cell (4): Cell Surface - Plasma Membrane and Primary Cell Wall

Previous plates have shown sections of plant cells. Here freeze etching and shadow casting of cells and isolated walls are used to obtain views of surfaces.

Plate 4a This face view of part of the plasma membrane (upper left) and cell wall (lower right) of a cell in an *Asparagus* leaf was obtained by the freeze-etching technique. Many particles about 10nm in diameter lie in the plasma membrane, sometimes in rows (e.g. arrow). The particulate texture of some areas (asterisks) is different from the surrounding membrane. Aggregates of particles sometimes occur. At one place the whole thickness of the plasma membrane has ripped away to expose another membrane surface within the cytoplasm (star). The particles in the expanse of membrane are probably proteinaceous, while the smooth areas between the particles represent an internal surface composed of the hydrophobic tails of lipid molecules, exposed when the membrane fractured along its mid line. x95,000.

The rapid freezing of the specimen preserved the membrane particles at the particular positions they were in at the moment of freezing. In life, however, they are free to diffuse around in the plane of the membrane, unless they are tethered, for example to elements of the cytoskeleton inside the cell or to parts of the cell wall outside the cell. Indeed there is evidence for a class of membrane proteins that physically link the wall to the cytoplasm through the plasma membrane (Plate 16). In the case of proteins that are free to diffuse, aggregates of various kinds may arise, forming specialised regions of the membrane. Protein-protein interactions may give rise to supra-molecular complexes. Examples of these phenomena are shown in Plates 22 and 33.

Multiple layers of microfibrils can be seen in the cell wall. The scattered holes may be plasmodesmata (PD).

The primary cell wall (i.e. the wall of a cell that is still growing) is built of long fibrils of cellulose held in place and crosslinked by xyloglucans and embedded in a less structured gel of pectins to form an extendable three-dimensional structure. In this specimen the initial freezing process preserved the cell wall in its native state. No extractions were carried out, therefore the image shows all wall components, still embedded in ice and broken along a very irregular fracture line. The next two micrographs present primary cell wall preparations after extractions designed to expose the cellulose microfibrils on their own (4b) and then the cellulose microfibrils cross-linked by hemicellulose bridges (4c).

Plate 4b As a cell grows, surface expansion progressively distorts the patterns in which microfibrils were initially oriented. The cellulose microfibrils illustrated here by means of shadow casting are in pieces of primary cell wall from which wall matrix materials have been extracted (cf. the native walls seen in Plate 4a). Plate 4b is a fragment of cell wall viewed from the plasma membrane side. The most recently deposited microfibrils remain more or less in their original orientation (in the axis at right angles to the long axis of the cell). Microfibrils that were deposited earlier, i.e. those lying deeper within the wall, have become pulled into more and more oblique orientations by cell growth. Scattered plasmodesmata are present in the wall fragment, their sites circumscribed by curved microfibrils (PD). (elongating pith cell, *Ricinus*, x22,000).

Cellulose microfibrils consist of polymers of glucose molecules joined together in long chains by β1-4 links. Each microfibril contains many such chains in parallel, arranged in a crystalline array in the microfibril's core and less regularly arranged at its periphery. The cellulose component of the cell wall is structurally significant because of the high tensile strength of the microfibrils. Microfibril alignment determines (in large part) the mechanical properties of the wall, and hence the shape of the cell as it expands during growth. For example, an array of parallel microfibrils is not easily extended in the axis of the microfibrils but is easily extended at right angles to that axis. Thus important related topics are: how cellulose microfibrils are made at the outer face of the plasma membrane (Plate 33) and how their deposition in particular orientations is regulated (Plate 32).

Plate 4c Cell wall material extracted from onion bulbs was treated to remove most of the pectins and rapidly frozen. Deep-etching of the frozen sample removed vitrified water from around the molecular framework of the wall. The etched surface was then coated with carbon and plastic to make a replica, which was rotary shadowed to reveal the microfibrillar structure. In this micrograph microfibrils are arranged in parallel layers running predominantly in two directions at right angles to each other (arrows). Examination of stereo views shows that fibrils at the same depth in the wall run in approximately the same direction. Removal of the pectin component reveals fine links between the cellulose microfibrils. These represent hemicellulose, probably xyloglucan, chains that cross-link the cellulose microfibrils. Such cross bridges may be very important in regulating cell growth. Certain enzymes can sever them specifically, and this relaxation of the cross-linking may permit a brief period of local wall expansion before the lattice is enzymatically restored. Growing cells may be able to control where and when this happens in their walls, and, on a larger scale, plants may be able to control where and when it happens in multicellular tissues, thus defining zones of growth. x75,000.

4a kindly provided by H Richter and U. Sleytr, reproduced by permission from *Mikroskopie*, **26**, 329-346, 1970); 4b kindly provided by K. Muhlethaler, reproduced by permission from *Ber. Schweiz. Botan. Ges.* **60**, 614, 1950; and 4c kindly provided by M. McCann, B. Wells and K. Roberts, reproduced by permission from *J. Cell Sci.* **96**, 323-334, 1990.

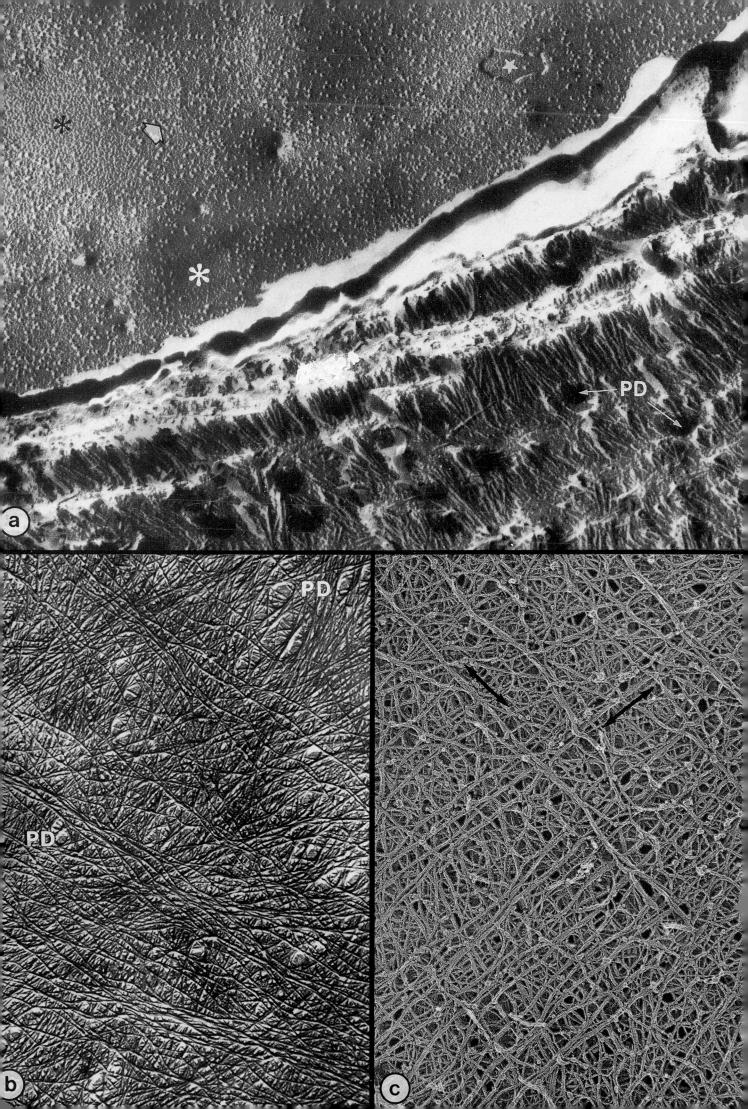

5   Nucleus (1): Nuclear Envelope and Chromatin

Introduction: The nuclear envelope is a specialised cisterna that surrounds the contents of the nucleus and controls the inward and outward flow of materials between the nucleus and the cytoplasm. Its two distinct membranes delimit an internal lumen, continuous with that of the endoplasmic reticulum through occasional connections. The outer membrane resembles endoplasmic reticulum in its ability to bind ribosomes. The inner membrane is linked to components of the nucleus, probably helping to organise them in three dimensions. A special category of cytoskeletal protein belonging to the *intermediate filament* family and known as *nuclear lamin* occurs in nuclei, with functions in organisation of the chromatin and its attachment to the nuclear envelope.

The nuclear envelope prevents mixing of nucleoplasm and cytoplasm, but also provides for transport in both directions. Proteins that are made in the cytoplasm and move into the nucleus include histones that bind to DNA and polymerase enzymes that catalyse DNA and RNA synthesis. Precursors of ribosomes (made in the nucleolus) and messenger RNA molecules are examples of macromolecules that move to the cytoplasm. These transport functions are accommodated by nuclear envelope pores. The envelope membranes are not themselves perforated, but are confluent around the pore margins. Each pore is founded on eight large protein complexes around its periphery, giving an octahedral outline. Additional proteins project across the pore from the octagonal side walls, reducing the effective diameter from an apparent 60nm to about 13nm. The pores are not entirely passive, but actively process some materials in transit - especially RNA strands exiting the nucleus.

Plate 5a The fracture in this freeze etched preparation of the nuclear region of a *Selaginella kraussiana* cell passed along the surface of the nuclear envelope (lower half of picture) and then broke through the inner (I) and outer (O) membranes. The upper half of the picture shows a relatively featureless fracture through the nucleoplasm (N). Nuclear envelope pores are seen in surface view (their helical pattern in this cell type is unusual), and in side view where the envelope is cross-fractured (arrows). Continuity between endoplasmic reticulum and nuclear envelope is seen (star). We thank B.W. Thair and A.B. Wardrop for this micrograph, reproduced by permission from *Planta,* **100**, 1-17, 1971. x24,000.

Plate 5b Where the nuclear surface is irregular, a single section can include both side (S) and face views of pores in the nuclear envelope (NE). Some face views (arrows) show a granule in the centre of the pore. Although regions of heterochromatin (C) touch the inner membrane, a halo of clear nucleoplasm lies around the pores (lower left). Cress root cell, x32,000.

Plate 5c The appearance of nuclear envelope pores depends on their position and orientation within the thickness of the section, which at this magnification (x100,000) is a slice about 5mm thick. Flat edges of the octagonal pore perimeter are visible in pores 1,3,4,8,11. Other features shown include: particulate components of the inner annulus (identifiable by proximity to chromatin - upper parts of pores 2,4) and outer annulus (identifiable from nearby polyribosomes - pores 5,9); fine fibrils traversing the pore lumen (pores 1,7,8,10) and apparently nearly occluding some pores (pores 4,9,11); pores arranged equidistant from a mass of chomatin (C) (e.g. pores 6,7,8,10); fibrils passing between chromatin and pore margins (arrows); polyribosomes (P) on the outer membrane; particles in the centre of some (1,2,4,6,7,8-11) but not all pores. *Vicia faba* root tip cell.

Plate 5d Side views of pore complexes are seen here, in the same material and at the same magnification as in 5c. The nucleoplasm is at the top of each picture. All pores show continuity of inner (1) and outer (O) envelope membranes at their margins. Other features vary, in part because the pores lie at slightly different levels with respect to the section thickness and in part because we may be seeing stages of movement of ribosome precursors to the cytoplasm. Units of the octagonal pore annulus are seen at the inner margin (e.g. single arrows) and outer margin (e.g. double arrows). Particles thought to be pre-ribosomal (large open arrows) are either: absent (top left); in the nucleoplasm near the pore (top centre); at the inner part of the pore lumen (top right); both inside and outside the pore (lower left); in the pore and just outside in the cytoplasm (lower centre); and in the pore as well as both inside and outside (lower right). These particles are smaller than mature cytoplasmic ribosomes, visible in the lower part of each micrograph. As in 5c, fine filamentous strands connect chromatin to the inner annulus.

Plate 5e If chromatin is placed on an aqueous surface, the surface tension spreads the constituent DNA strands. Rotary shadowing (see introduction) and electron microscopy can then reveal the strands, as here. The DNA is folded at intervals into beads, known as *nucleosomes*, by wrapping it around histone molecules. In the nucleus these beaded strands are packed into 30nm wide chromatin fibres, held in place by further histones, and these in turn are packed into higher order arrangements. The degree of condensation of chromatin reflects the genetic activity of the cell and the stage of the cell division cycle. Electron micrographs of ultra-thin sections usually show patches of dense *heterochromatin* and more dispersed *euchromatin* regions. Euchromatin is regarded as being the more genetically active. Specialised cells often show much heterochromatin (e.g. Plate 29), whereas cells that are transcribing large proportions of their genes may show a preponderance of euchromatin (e.g. Plate 2). Plate 54 contains a striking comparison of two extremes of chromatin condensation. Micrograph (x87,000) kindly provided by R. Deltour, reproduced by permission from *J. Cell Sci.* **75**, 45-83, 1985.

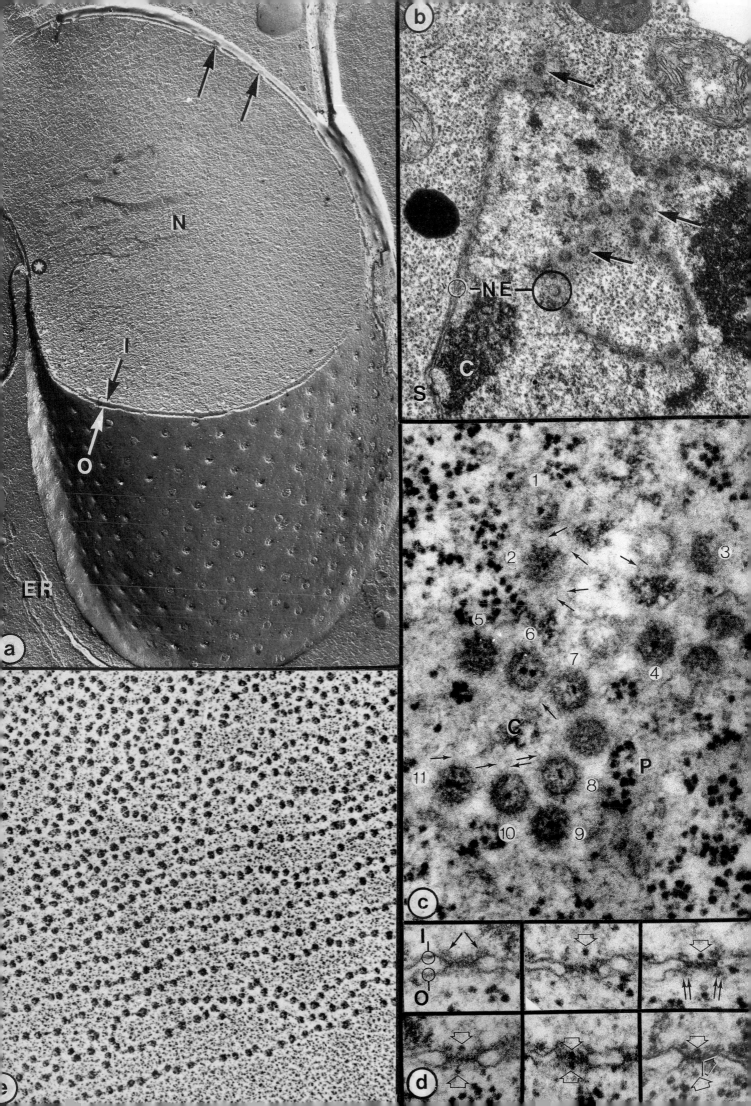

6   Nucleus (2): Nucleolus

Introduction: The nucleolus is a characteristic feature of the nuclei of eukaryotic cells. Major constituents are a repeated sequence of genes that code for the RNA of ribosomes, and a mass of the products of their activity. Most of the RNA in the nucleolus is a precursor of the RNA of ribosomes, and the dynamic processes observable in nucleoli centre on the synthesis of this material, first in fibrillar form and later its association with ribosomal proteins (imported from the cytoplasm) to form ribonucleoprotein particles. These *pre-ribosomes* are then transported from the nucleus to the cytoplasm, undergoing maturation steps as they pass through the nuclear envelope pores.

Plate 6a-f   These six micrographs illustrate dynamic processes in a nucleus of a living tobacco cell growing in artificial culture. The nucleolus (centre of each picture) is surrounded by a bright halo, due to the phase-contrast optical system used. At the start of the sequence (a) the nucleus (margin outlined by arrows) is ellipsoidal. The nucleolus, about 7μm in diameter, contains a large nucleolar void about 70μm^3 in volume. One minute later (b) the void is connected to (arrow), and apparently emptying into, the nucleoplasm. After another 15 seconds (c) the void is much smaller. It is just detectable after a further 15 seconds (d), but cannot be seen (e) 2 minutes after the first picture was taken. By this time the whole nucleolus has shrunk by about the volume of the void. One hour later (f) another void has formed in the nucleolus, which has regained its original total volume, and the nucleus itself has become more rounded. The inference from these time lapse pictures is that the nucleolus is accumulating a product that is intermittently delivered to the nucleoplasm. The mechanism of rapid contraction of the void is not understood. No cell membranes are involved. All x1,400. Kindly provided by J.M. Johnson, reproduced by permission from *Amer. J. Bot.*, **54**, 189-198, 1967.

Plate 6g   Part of a nucleolus in a *Vicia faba* root tip cell nucleus is shown here magnified x58,000 (see Plate 1 for the same material in a light micrograph). A central void (V) lies within the nucleolus (as in a, f). Granular (G) and fibrillar (F) zones constitute the dense material, along with small areas of chromatin (arrows) ramifying through electron-transparent channels. Condensed (CC) and dispersed (DC) chromatin is seen in the nucleoplasm; the former at one point touching and penetrating into the nucleolus (dashed lines). This chromatin is the *nucleolar organiser* region of the genome, shown again in condensed form in Plate 43. The void contains scattered fibrils and particles. Presumably they are discharged into the nucleoplasm when the void pulsates.

Plate 6h   This high magnification (x150,000) view shows 6-8nm nucleolar fibrils (F) and 12-14nm granules (G) near the periphery of a nucleolus (*Vicia faba* root tip). The granules are smaller than cytoplasmic ribosomes,

consistent with them being a precursor form. Some granules are attached to or associated with fibrils. Convoluted fibrils of DNA-histone are seen in the chromatin (C), near which lies a perichromatin granule (arrow), with angular profile, 50-60nm in diameter. These populations of granules probably represent stages of maturation of ribosomes and messenger-RNA transcripts, prior to their export to the cytoplasm.

Plate 6i   Plate 5e demonstrates fine details of fibrillar material that are not visible in ultra-thin sections. There chromatin was spread out and stained for electron microscopy. The same method is applied here to the nucleolus, where the spreading procedure uncovers the macromolecular organisation of ribosomal genes and transcripts.

There is one DNA fibre in each nucleolar organiser (of which there may be several in a nucleus). It is a chain of thousands of copies of the genes for ribosomal RNA (17,000 in pea), each separated by intergene spacer regions (arrows). The position of each gene is revealed by its attached gene products, i.e. transcripts of ribosomal RNA (example between brackets). These are visible as branches, shortest where they have just started to be formed by an RNA polymerase enzyme molecule lying at the base of each branch, and longest where the polymerase has travelled right along the gene, transcribing the product as it moved. The gradient of branch length represents a snapshot of the numerous RNA polymerases at work on each gene, their activities halted just as they were at the instant the preparation was made. It can be seen that after a period of transcription, each projecting RNA strand starts to fold and interact with granular material, seen as terminal knobs. This is interpreted as the first stage of association of pre-ribosomal RNA with ribosomal proteins.

There are in fact two size classes of large RNA molecule in each ribosome, plus two much smaller RNA molecules, and about 80 different proteins, all arranged in one large and one smaller ribosomal subunit. The total assembly process is therefore very complex. It is also, in total, very rapid. Plate 2 (and many others) illustrate the vast numbers of ribosomes in a cell. The numbers have to be approximately doubled every time the cell divides. Moreover the rate of production must also take account of breakdown ("turnover") of cytoplasmic ribosomes. Estimates of the required rates are in the tens of thousands per minute, per cell. This accounts for the presence of thousands of genes for ribosomal RNA in each nucleus, each being transcribed by tens of RNA polymerase molecules at any given instant, and for the evidence, in the form of pulsation of nucleolar voids in living cells (a-f), of intense productivity and export to the cytoplasm. Kindly provided by R. Greimers and R. Deltour, reproduced by permission from *Biol. Cell.* **50**, 237-246, 1984. x38,000.

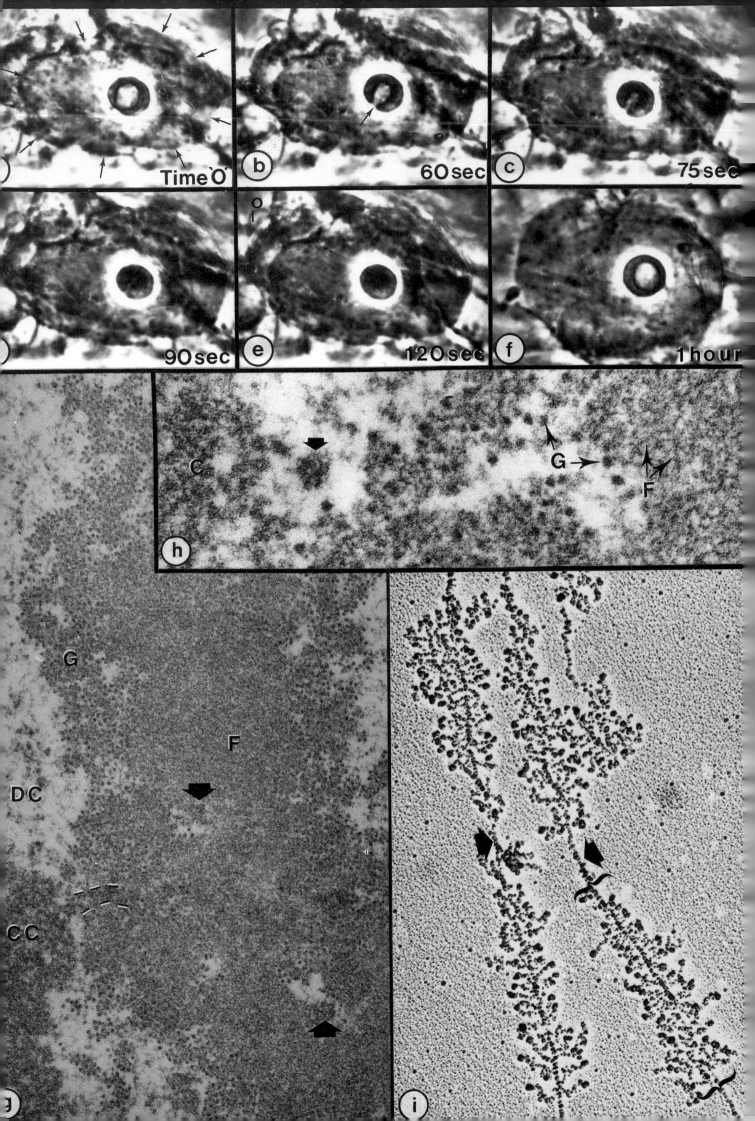

Time 'O'

b 60 sec

c 75 sec

90 sec

e 120 sec

f 1 hour

h

C G F

G

F

DC

CC

i

Plates 7-14 illustrate the endomembrane system of plant cells, a functional continuum that starts with the endoplasmic reticulum and leads through the Golgi apparatus to the cell surface or vacuoles. Plates 7-9 show the endoplasmic reticulum and Plates 10-14 the Golgi apparatus, including diversity of structure and examples of function.

7 Endoplasmic Reticulum and Polyribosomes

Introduction: The endoplasmic reticulum (ER) membrane system is probably present in all plant cells. It is very dynamic and variable in its quantity, structure and organisation, and in its position in the cell. It is multifunctional and may be locally differentiated within a cell, with diverse biosynthetic and regulatory roles. Plates 7 and 8 show extreme examples of rough and smooth ER, Plates 8 and 36 a special form known as cortical ER, Plate 36 the dynamic alterations of the ER during the cell division cycle, and Plate 54 striking regional differentiation of the ER in a single cell.

Plate 7a This unusually regular arrangement of the ER is found in surface glands (trichomes) on leaves of *Coleus blumei*. The 12 to 14 parallel cisternae are closely stacked, confining the intervening cytoplasm to very thin layers, which, however, expand in some places (stars). The cisternae interconnect at branch points (arrows) and at swollen intracisternal spaces (S), which contain diffusely flocculent material, perhaps the product of the system. The product (Pr) secreted by the gland passes through the inner layer of the cell wall (CW) and accumulates beneath the cuticle (C). The ER is continuous with the outer membrane of the nuclear envelope (open arrow, lower right). The plastid stroma (P) is filled with a material which (after specimen preparation) is very dense to electrons. x22,000.

Plate 7b One of the spaces into which the cisternae open is shown here enlarged from (a) The tripartite substructure of the ER membranes (black and white arrow) is thinner than the adjacent tripartite plasma membrane (black arrow). The lumen (Ci) of the ER is not normally connected to the space outside the plasma membrane (Co) in any cell, but in this case cisternal contents may have been released to the exterior either as a transient normal process, or as an artefact of specimen preparation. Similarity in size and flocculent contents between Ci and Co suggests that the ER could have discharged to the exterior by rupturing the narrow bridge of cytoplasm that separates Ci and Co, leaving a fragment (open arrow) of tripartite plasma membrane enclosing several ribosomes. In the vast majority of cases the ER does not fuse with other membranes. Rather it gives rise to vesicles which serve as intermediates and which *are* equipped to fuse with other systems, especially the Golgi apparatus (see Plate 10). x75,000.

Plate 7c The main difference between rough and smooth ER is that the former has ribosomes attached to its cytoplasmic surface. Neither the ribosomes nor the membrane are special in rough ER. The characteristics of the proteins that are being made determine whether ribosomes attach (indirectly) to the membrane. Ribosomes are part of the molecular machinery for synthesising protein molecules, and polyribosomes are chains of ribosomes which are linked by evenly-spaced attachment to a strand of messenger-RNA carrying the nucleotide triplet specification of the protein's sequence of amino acids. Proteins that are destined for export from the cell or for incorporation in membranes are made on rough ER. The first part of their amino acid chain is a *signal sequence*. It binds to a *signal recognition protein* which then in its turn is able to attach to *docking protein* in the ER membrane. In such cases the protein molecule that emerges from each ribosome as it translates the mRNA message is injected into the ER membrane. It either stays there when it is completed or else passes into the ER lumen (see also Plate 14). Proteins that stay in the cytosol (at least initially) lack the signal sequence, so when they are being made, the signal-recognition-docking proteins do not bind them to the ER.

7c shows oblique surface views of cisternae of rough ER in a hair cell of the alga *Bulbochaete,* carrying hairpin-shaped polyribosomes. Polyribosomes also lie on the outer surface of the nuclear envelope (N-nucleus). The longest examples include about 20 ribosomes in a total length of 500nm. A messenger RNA molecule of this length would contain about 1500 nucleotides, with each set of three coding for one amino acid and occupying a length of one nm. Hence the (unknown) protein that is being made in this particular cell is likely to have about 500 amino acids and a molecular weight of about 60,000 (taking the average molecular weight of one amino acid to be 120). x67,000. (We thank T. Fraser for this micrograph).

The inset shows a free cytoplasmic polyribosome configuration consisting of a tight helix of ribosomes unattached to membranes. The helix is longer than the section is thick, so the 0.4µm length that is included here may not represent the total length. Comparable polyribosomes in liver cells are based on long strands of helically wound mRNA. Bean root tip cell, x90,000.

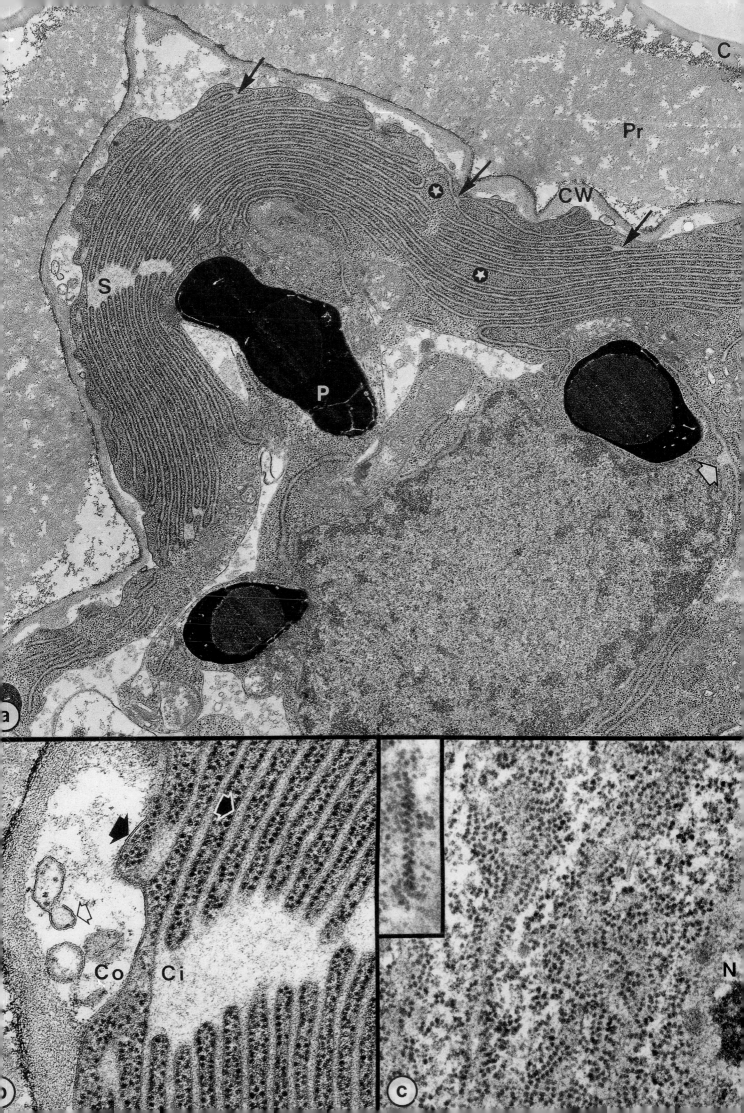

8   Smooth and Cortical Endoplasmic Reticulum

Introduction: Plants manufacture and secrete an enormous diversity of chemical substances. In the example illustrated in (a-c), a white floury material, or *farina*, is produced by single celled glands developed particularly by certain members of the flowering plant family Primulaceae. Farina glands are one of several categories of plant gland in which a system of smooth ER membranes is exceptionally well developed. Smooth ER is seldom seen in such massive arrays, but most cells possess a distinctive form of (largely) smooth ER lying very close to the inner face of the plasma membrane. This *cortical ER* is also shown in electron micrographs in Plates 10a and 53c, and by confocal microscopy in this Plate (d), and in Plates 16 and 36.

Plate 8a and b These scanning electron micrographs show the floury appearance of the surface of an *Auricula* sepal, visible to the eye as a yellow powder, and at low magnifications ((a), x1,100) as mounds of fine crystals. Each mound ((b), x5,800) is a stalked gland cell, covered by its secretion product in the form of contorted ribbon-shaped or linear crystals radiating from the cell surface. A mixture of substances is present, the major components being flavonoids. Some people are allergic to the farina.

Plate 8c In the course of preparation for ultra-thin sectioning, this farina gland on a young petal of *Primula kewensis,* x40,000) had its secretion product dissolved away. Consequently none is seen outside the cell wall (CW). The main feature of the cytoplasm is the system of smooth ER tubules, each one 60-100nm in diameter, ramifying through the cytoplasm. Not many branched tubules are visible (arrows), and it is obvious from the varied profiles displayed in the section that the tubules do not lie in a regular pattern. The ground substance of the cytoplasm is granular, and contains a rather sparse population of ribosomes, usually in clusters, presumably polyribosomes (e.g. circle). The mitochondria (M), being nearly circular in outline, are probably nearly spherical, but the electron-transparent areas within them suggest that they might have become swollen during chemical fixation. Microbodies (MB), with granular contents and single bounding membrane, are conspicuous. The plastids (P) are small and undifferentiated. An electron-dense deposit lines parts of the tonoplast of several of the vacuoles (V). In a number of places (asterisks) the configuration of the tonoplast suggests that material such as the small droplets of electron-dense material that are present throughout the cytoplasm (especially at lower left) associated with the ER, may be being incorporated into the vacuoles.

The significance of this cytoplasmic organization in relation to the synthetic and secretory activity of the farina gland is obscure. The plasma membrane (PM) is fairly smooth, and there is no evidence that products of secretion leave the cytoplasm in the form of vesicles by reverse pinocytosis (*granulocrine* secretion). The outward transport is more likely to be *eccrine*, that is, by a flux of free molecules (not in vesicles) across the plasma membrane and thence to the exterior by diffusion through the cell wall. Crystallization occurs upon exposure to the air.

The role of the labyrinths of smooth ER is not understood. Smooth ER is present in a variety of plant glands secreting fats, oils, and fragrant essential oils, and is also abundant in epidermal cells that are making lipid molecules to be deposited in the external cell wall - waxes, cutin and suberin. In general the ER has been found to possess the enzymes necessary for making complex lipids, given the ingredients of fatty acids (made in plastids) and lipid head groups (made by cytoplasmic enzymes). These raw materials come together in the cytoplasmic face of the ER membrane to produce a variety of products. Depending on their nature, they may accumulate until they form lipid droplets, initially still in the membrane but then liberated into the cytoplasm for storage or transport around the cell. Special enzymes can "flip" lipids from the cytoplasmic face of the ER membrane, where they are made, to the luminal face, correcting the imbalance that arises from asymmetrical synthesis. In this way the ER membrane grows in surface area. Expanses of membrane may then be mobilised to other systems in the form of vesicles- especially to the Golgi apparatus and from there to the plasma membrane. Another option is that the products of lipid synthesis can be picked out of the cytoplasmic face of the ER by lipid transfer enzymes and moved as individual molecules to sites of utilisation.

Plate 8d Time-lapse pictures of two small parts of the cell cortex of a living cell were taken at one minute intervals by confocal laser scanning microscopy. The ER and some other membrane systems were vitally stained with the fluorescent dye rhodamine-123. It is very common to find a network of ER channels in the cell cortex and to see them constantly moving and forming and breaking interconnections. Times (minutes) are as shown. Arrows point to areas of marked change (upper panel). Sometimes new channels form in straight lines (arrows, lower panel) as if they were extending alongside microtubules or actin microfilaments (see Plate 31b). Often they link with flatter cisternae (upper panel and Plate 36). Suspension culture, tobacco BY-2 cell, x240 (upper), x460 (lower).

The function of this cortical ER is not understood. It is strategically located to act in concert with the plasma membrane in localised ion-pumping functions, for example regulating the level of calcium ions in the cytoplasm. The ER and the plasma membrane do not fuse. However, as demonstrated in Plate 16, it is strongly linked through the plasma membrane to the cell wall - strongly enough to remain there when the cell is plasmolysed.

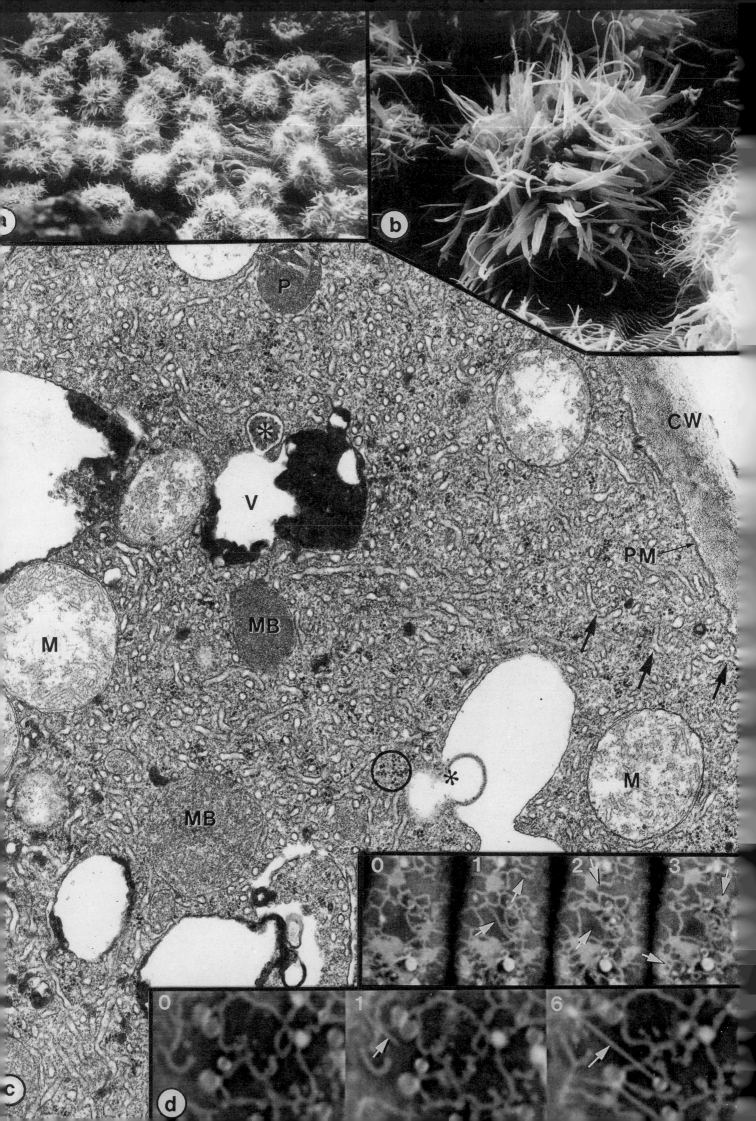

a

b

P

*

V

CW

M

MB

PM

MB

M

*

○

0   1   2   3

0   1   6

c

d

9   Membranes of the Golgi Apparatus

Introduction: Golgi bodies are stacks of circular, flattened cisternae, each delimited by a single membrane. The contents, biochemical composition and functional activity of the cisternae change from one face of the stack to the other. In cells of animals and lower plants, one outer cisterna is often associated with nearby endoplasmic reticulum (ER). This is referred to as the *cis*-compartment, at the *cis*-face of the stack. Large secretory vesicles arise at the opposite cisternae, at the *trans*-face. Between the *cis*- and *trans*- regions are medial-cisternae. An irregular *trans*-Golgi network of loosely-associated membranes lies near the *trans*-face of the stack. The successive compartments have different biosynthetic capabilities, operating in sequence.

The Golgi apparatus processes macromolecules and packages them for export to the cell surface or vacuoles. The products include lipids and proteins, carried in the membranes of the secretory vesicles, and cisternal contents, received from the ER by transitional vesicles or synthesised *de novo* within the Golgi apparatus. Often they are protein molecules bearing short sugar side chains i.e. glycoproteins. Most glycoproteins enter from the ER as proteins bearing a standard form of side chain (an *N-linked glycan*) which is first shortened and then extended again by sugar transferases in *cis*- and medial cisternae and completed at the *trans*-face and *trans*-Golgi network to form mature glycoproteins (Plate 13).

While the main activity of the Golgi apparatus of animal cells is processing of luminal and membrane glycoproteins, the evolution of cell walls in the progenitors of plants wrought a major change in emphasis in the biosynthetic activity of plant Golgi bodies. As well as glycoproteins, they make pectins, hemicelluloses (xyloglucans), mucilages, and other polysaccharide secretory products and components of cell walls (Plates 10-13). Unlike glycoprotein biosynthesis, this is done independently of import from the ER, although the necessary *enzymes* in the Golgi membranes must be imported from the ER.

It is not clear how plant Golgi stacks function. There is no evidence that successive cisternae deliver their contents to one another by establishing direct membrane continuity. As in animal cells, contents may be selectively moved from one cisterna to the next by small vesicular shuttles operating at the margins of the successive compartments. However there is also evidence that whole cisternae can move progressively through the *trans*-region and eventually leave the stack.

Plate 9a   The membranes of this freeze-etched Golgi body were exposed by a fracture which passed along an expanse of membrane (upper right), through a collection of vesicles, then steeply down in a succession of narrow steps through 6-8 cisternae, and finally along a more level plane where it exposed face views of a cisterna (centre) and numerous vesicles.

Comparisons with ultra-thin sections from the same material (cells of the green alga *Micrasterias*) allow identification of the small vesicles at the upper right as transitional vesicles (TV) entering the *cis*-face of the Golgi body. The first cisterna in the stack is pierced by a cleft (arrow), and had the cell not been killed, it is probable that the split would have continued through the remaining cisternae. In other words, this cleft may be the first sign of a fission process which would have resulted in the production of two Golgi bodies from one. This is probably a general mechanism for multiplication of Golgi stacks.

The larger vesicles (V) at the bottom and lower left of the micrograph represent the membrane-bound products formed at the *trans*- face of the Golgi body. Other small vesicles, possibly intercisternal shuttles, lie at the cisternal margins. The extensive surface labelled M probably represents the interior of the membrane of a *trans* cisterna. Numerous particles (about 7,000 per μm^2) lie in the membrane at the central region of the cisterna, declining markedly in frequency at the vesiculating periphery. Vesicles at the maturing face (V) have relatively few particles. x82,000. Kindly provided by A. Staehelin, reproduced by permission from *J Cell Sci.*, 7, 787-792, 1970.

Plate 9b and c   The *cis*-face of this Golgi stack (in another green alga, *Bulbochaete*) is uppermost, where the bifacial ER, transitional vesicles (TV), and the first, highly-fenestrated, *cis*-cisterna (arrow) can be seen. The transitional vesicles are "coated" (see Plate 15). Vesicles with visible contents are present at the margins of the stack and near the *trans*-face (V). Six cisternae have granular contents, both in the central regions and at their periphery, but in the three cisternae closest to the *trans*-face the central regions are clearly tripartite (see (c), a higher magnification view of the same Golgi stack), and the intra-cisternal compartment is very thin, thus restricting the contents to the margins, where vesicles are formed. In the most mature cisternae the membranes have moved apart again, and the former contents have been packaged into vesicles. The central membranes may break down into fragments (F).

In the absence of labelling procedures to identify enzymes and products in different regions of the Golgi bodies shown in this Plate, it is impossible to assign successive cisternae in the stacks to *cis*-, medial- and *trans*- compartments. Proximity to ER identifies the first cisterna at the *cis*-face. The three cisternae with distinctive tripartite membranes in (b) and (c) probably represent the *trans*-compartment, and the very irregular fragments of membrane and vesicle around "F" in (b) are part of a *trans*-Golgi network. Subsequent plates analyse the stratification of the stacks in more detail. Magnifications: b, x70,000; c, x155,000, (micrographs kindly provided by T.W. Fraser).

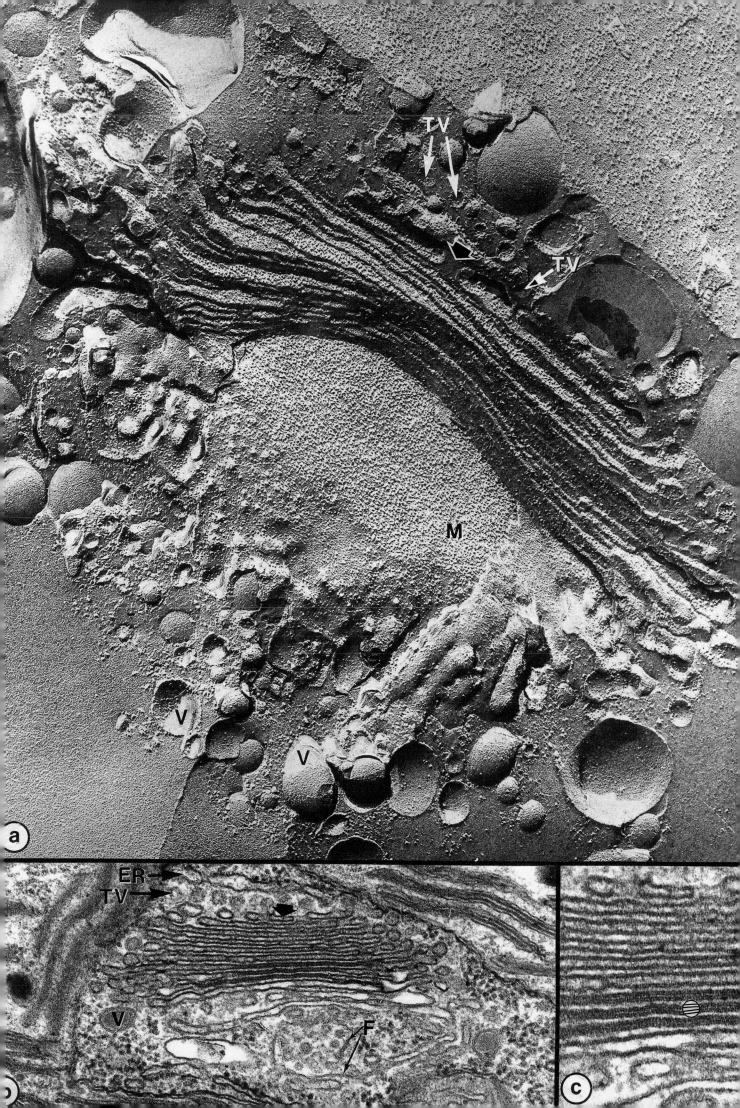

10 Relationships between Golgi Apparatus, Endoplasmic Reticulum and Nuclear Envelope

Introduction: *Functional* relationships between the ER (including the nuclear envelope) and the Golgi apparatus are especially evident in certain algal cells, where the two systems enter into consistent *positional* relationships. Micrographs of these cells are therefore illustrated here to aid understanding of the combined operations of the ER - Golgi continuum. The same functional relationship between the ER (but not the nuclear envelope) and Golgi apparatus recurs in higher plants, though the positional relationships are much less obvious.

Plate 10a The filamentous green alga *Bulbochaete* possesses hair cells (chaetae) which make a mucilage and contain a conspicuous Golgi apparatus consisting of numerous discrete stacks (e.g. G 1-3), between which cisternae of rough endoplasmic reticulum (ER) ramify. The Golgi stacks are oriented with respect to the subjacent cisterna of ER only, thus in G-1 the forming face is lowermost, in G-2 uppermost, and in G-3 at the left hand side.

The ER is bifacial where it underlies the Golgi stacks. It lacks ribosomes in the zones (between arrows at G-2 for example) nearest to the *cis*-faces of the stacks, presumably because numerous transitional vesicles (TV) are formed at these zones. The opposed face of the ER bears ribosomes as usual. The inset at lower right is part of a section adjacent to that used for the main picture, and shows Golgi stack G-2. Whereas G-2 (main picture) merely shows transitional vesicles (bearing a fuzzy coat on their membrane), G-2 (inset) shows what can best be interpreted as a stage of formation of a coated transitional vesicle (upper arrow) and a stage of coalescence of another such vesicle (lower arrow) with the first *cis*-Golgi cisterna.

Numerous other vesicles occur in the cytoplasm. The larger type (V) is more or less spherical, with granulo-fibrillar contents, the product of the Golgi apparatus. These are formed at the margins of cisternae at the *trans*-face of the Golgi stacks. As shown in the inset at top right, their bounding membrane is tripartite (arrow), and of the same staining characteristics and dimensions as seen in the plasma membrane, with which they fuse to release their contents to the exterior.

As with Plate 9, it is not possible to discriminate between the *cis*-, medial- and *trans*- Golgi compartments within the stacks. However, the very numerous coated vesicles at the margins of the cisternae (distinct from the larger vesicles (V) that contain the secretory product) must represent the vesicles that shuttle intermediates and membranes between the successive compartments.

The ER is again bifacial where it lies alongside the vacuole (Va, between arrows). Hair cells in *Bulbochaete* have no chloroplasts (in fact they are very unusual in the plant kingdom in not containing plastids of any kind), yet they produce abundant Golgi vesicles, and clearly require

raw materials. It may be that the latter enter the hair cell *via* plasmodesmata (PD) from photosynthetic cells on the other side of the cell wall (CW). The vacuole could be a storage reservoir for raw materials, and the closely juxtaposed tonoplast (T) and bifacial ER a route for entry into the secretory endomembrane system. Once in the ER, substrates and intermediates could pass to the bifacial regions subjacent to Golgi stacks. Main picture and lower inset x52,000; upper inset x120,000. Kindly provided by T.W. Fraser).

The nature of the secretory product in *Bulbochaete* hair cells is not known. However it is likely to have a protein component, judging by the clear structural evidence for input to the Golgi apparatus from the ER. In general the ER is involved in two stages of glycoprotein synthesis. First the protein backbone is made by polyribosomes linked to the ER membrane by signal-recognition and docking proteins. In the present case the product enters the lumen and does not stay in the membrane (Plate 7c shows the polyribosomes in question, consistent with a molecular weight of about 60,000 for the protein). Several categories of oligosaccharide side chains are found in plant glycoproteins. One of the commonest attaches to asparagine residues in the protein. The second stage in synthesis of this form of glycoprotein is addition of a standard form of side chain containing 14 sugar units, donated from a carrier enzyme in the ER membrane. This happens as soon as the asparagine residues enter the ER lumen. After preliminary trimming in the ER and transfer to the Golgi apparatus in transitional vesicles, some sugar units are trimmed off and others are added in specific sequences in successive Golgi compartments.

Plate 10b In many algae, as here in *Tribonema*, the Golgi bodies lie alongside the nucleus, and transitional vesicles (TV) are seen between the nuclear envelope (NE) and the *cis*-face. The ER and the outer membrane of the nuclear envelope are often continuous, even in higher plants, giving luminal continuity between the two systems (see Plates 5, 7). Indeed the nuclear envelope is often regarded as a specialised cisterna of ER. However in higher plants the nuclear envelope does not (or not obviously) serve as a source of precursor transitional vesicles for the Golgi apparatus, as it does quite frequently in algae. The examples in the micrograph are used to illustrate presumed stages in formation of coated transitional vesicles, as in the sequence labelled 1-7. The region of the nuclear envelope that produces vesicles has a thick (up to 100nm) layer of grey-staining fibrillar material on its outer surface (e.g. star). This is clearly significant for vesicle production since it is not seen elsewhere on the nuclear envelope. Further details of coated membranes and coated vesicles are given in Plate 15. x60,000. Kindly provided by G.F. Leedale, reproduced by permission from *Brit. Phyc. Jour.*, **4**, 159-180, 1969.

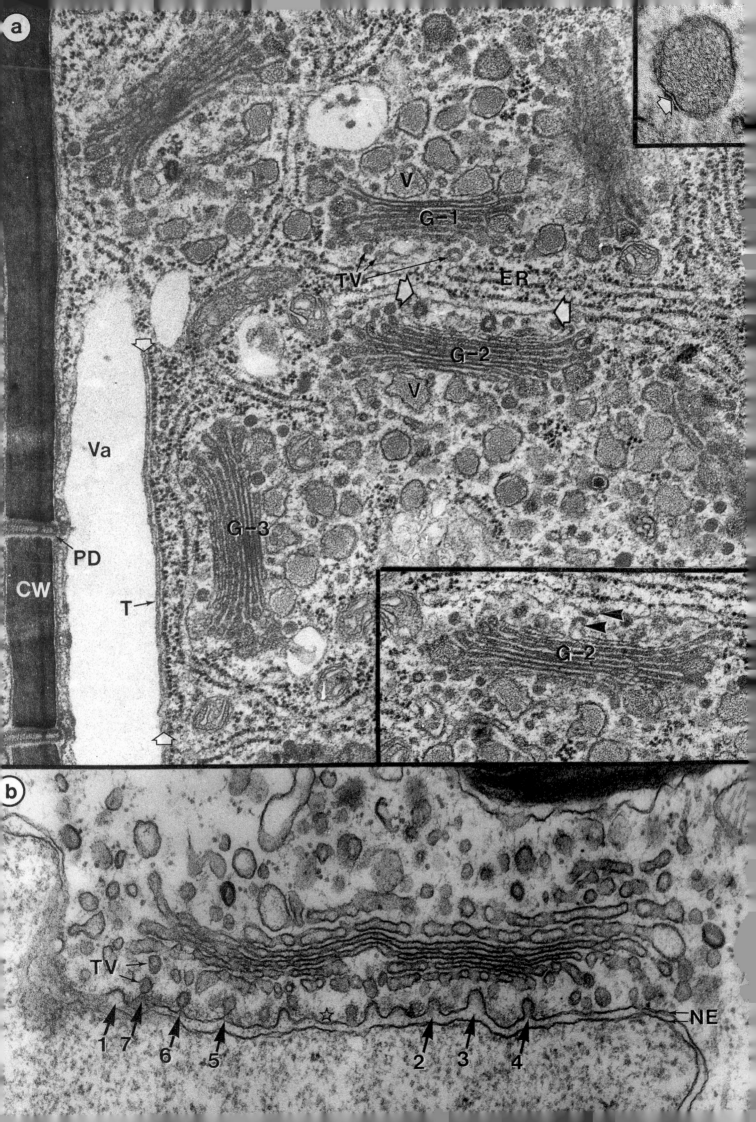

11 A Visual Example of Biosynthesis in the Golgi Apparatus: Production and Exocytosis of Scales

Introduction: These micrographs use a genus of single-celled planktonic flagellates in the Haptophyceae to illustrate central principles of Golgi apparatus function. (a)-(f) depict the end product of the system and (g)-(h) the specificity and mode of its formation. *Chrysochromulina* is the most important of the marine nanoplankton: tiny organisms at the base of marine food chains. Like many other scaly flagellates, it covers its otherwise naked protoplast with sculptured scales. In this genus the scales are predominantly carbohydrate, but in others the carbohydrate serves as a foundation for calcite, to form the coccoliths of the geologically important Coccolithophorids, or silica, to form the silicified scales of members of the Chrysophyceae. The scales are visually distinctive and can be recognised in ultra-thin sections in their final destination outside the cells, and also during development within Golgi cisternae.

Plate 11a A discarded scale case of *Chrysochromulina pringsheimii* is seen here in a shadow-cast whole mount. Two types of scale are visible - long pins at the ends of the case, and smaller scales, of which a few have broken away from the case (arrow).

Plate 11b-d (b) is an ultra-thin section of *C. chiton*. Although the scale case in incomplete, two types of scale are easily distinguished. Outer scales (O) have a flat base and a curved rim. A shadow-cast example is shown in (d), with the pattern of fibrils on the inner face (closest to the plasma membrane) exposed to view. Inner scales (I), lying between the outer scales and the cell itself, are slightly curved with a small recurved rim. In (c) they are viewed from the inside (upper) and also from the outside (lower). Their fibrils are exposed on the inner face but are overlain with amorphous material on the outer face. The shadow-cast outer face shows the recurved rim, formed from circumferential fibrils. (b) also shows nucleus, N; chloroplast, C, with stalked pyrenoid, P; mitochondria, M; and the single Golgi stack, G, with the *trans*-face uppermost.

Plate 11e, f These two sections illustrate genetic specificity in the shape of the scales. (e) is genetically deviant and differs from (b) in the shape of the outer scales (O), which have the same type of base, but lack the curved rim. The inner scales (I) are normal. (f) shows outer (O) and inner (I) scales of *C. camella*, in which the outer scales are shaped like cups with four rings of perforations on the sides. Numerous other distinctive species-specific forms could be illustrated.

Plate 11g, h Comparisons between the normal form (g) of *C. chiton* (scales as in (b) (c) and (d)), and the deviant form (h) (scales as in (e)) highlight four features of Golgi structure and function in this system:-

(i) *Golgi stacks are polarised*: ER and transitional vesicles (between open arrows) at the *cis*-faces are at the lower part of each picture.

(ii) *Scales are formed in Golgi cisternae*: the first visible signs of scales are indicated by the arrows. Recognizable outer (O) and inner (I) scales are present, each type within its own cisterna. The example of a normal outer scale in (g) shows that the shape of the scale and the cisternal membrane match each other, like a casting in a mould (arrowheads). Cisternae at the *trans* face, close to the plasma membrane (PM), contain scales of mature appearance. Liberated scales are shown in (h).

(iii) *Cisternae are dorsiventral*: the scales in the successive cisternae all develop in the same orientation, with their outer faces towards the *trans*-face.

(iv) *Genetic specificity is expressed in the Golgi apparatus*: the specific character which distinguishes the two forms of *C. chiton*, i.e. the shape of the scales, is seen to be developed within the Golgi cisternae.

Scale production thus illustrates formation of highly distinctive species-specific structures by the Golgi apparatus. The cisternae are shown to possess individual synthetic capabilities, and, like the Golgi stack as a whole, to be dorsiventral. These attributes must be founded in the spatial distribution of enzymes in or on the membranes of the Golgi cisternae. Other, even more complex examples, are known, for example, production of more than one type of scale *within a single cisterna* - a phenomenon that has its counterpart in a few situations in fungi and higher plants, where different kinds of product can be detected in a single cisterna and yet become sorted into discrete vesicles.

Despite the visual evidence for input from the ER and stagewise manufacture of distinctive products, we do not know how the system works. The stacks have an extensive *cis*-region of uniform cisternae, then there is a region containing a few cisternae with pronounced dilations (shown especially clearly in (h)). This could be a medial-Golgi compartment. The scales actually make their appearance in the region lying *trans*- from the dilated central cisternae. This could be the *trans*-compartment, perhaps along with an extensive *trans*-Golgi network, and the scales may still be maturing while the cisternae are moving to the cell surface. Are successive compartments static and interlinked by shuttle vesicles as in the animal model? Scales are not seen (and would not fit) in small vesicles, so this may be a case where whole cisternae flow through the system in the *trans*-direction. Alternatively the stack may be static in *cis* and medial zones, and cisternal flow and maturation limited to the *trans*-Golgi region.

(a) x1,750; (b) x10,000; (c)-(h) all x30,000. Kindly provided by Professor I. Manton. (a) reproduced by permission from *Jour. Marine Biol. Assoc. U.K.*, **42**, 391, 1962; (b,c,e) from *Jour. Cell Sci.*, **2**, 411, 1967; (f) from *Arch. Mikrobiol.* **68**, 116, 1969; (h) from *Jour. Cell Sci.*, **2**, 265, 1967.

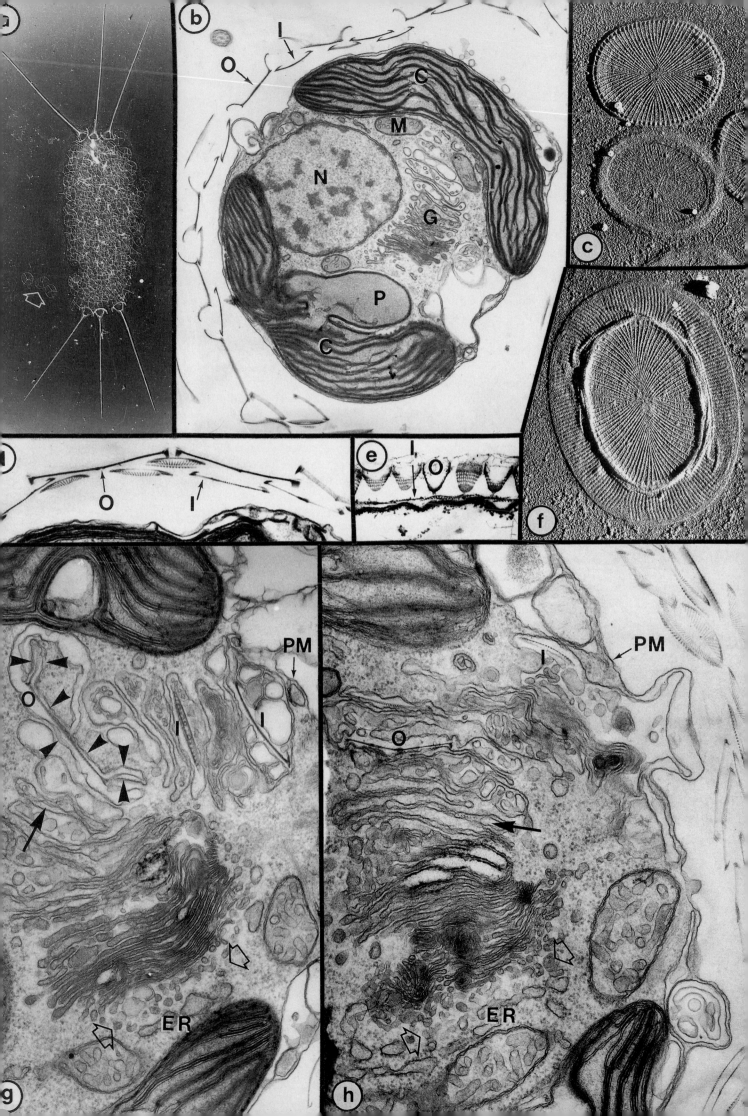

12   Mucilage Secretion by Root Cap Cells

Introduction: The Golgi apparatus of root cap cells produces mucilages that are exported by secretion through the plasma membrane and extruded through cell walls to the surface of the root tip. The phenomenon has provided much information on the properties of plant Golgi systems and is also very important in the life of the plant. Functions include lubrication as the root tip grows through the soil, protection against pathogens, altering the physical properties of soil, and providing a persistent "mucigel" at the surface of the root that harbours a variety of beneficial microorganisms accompanied by sloughed-off but still living root cap cells.

Remarkable quantities and rates of production are involved. Each root cap cell may have several hundred Golgi bodies, which behave as a synchronous population in the cytoplasm. Each root cap may slough off hundreds or thousands of still-secreting cells every day (in maize the number is estimated to be some 12,000, and a single plant has hundreds of root tips). The "slime" expands many-fold in volume when it hydrates in the soil, so the total volume from a crop of plants is staggering (more than $1000m^3$ per hectare in an annual growth cycle of maize plants). At the level of cell biology, the combined cells of one maize root cap have some 5 million Golgi stacks and produce 2 million vesicles per minute. When these fuse with the plasma membrane to secrete their contents, they augment the surface area of the plasma membrane by about 10% of its own area per minute. Compensating processes must remove plasma membrane at an equivalent rate to maintain a steady surface area and to recycle the constituents back to the ER at the beginning of the endomembrane continuum. Clearly, static micrographs cannot do full justice to this highly dynamic system.

Two grasses - timothy (*Phleum*) and maize (*Zea*) - are used in this plate because they illustrate different types of vesicle formation.

Plate 12a The periphery of this root cap cell of *Phleum pratense* shows part of the nucleus (N) and the thick, mucilage-laden cell wall (CW). The cytoplasm contains numerous Golgi bodies (G) whose cisternae contain fibrillar material, visible even close to the *cis*-face (C). The cisternae remain flat, with small vesicles at their periphery, but they increase in thickness (see numbered sequence in the Golgi stack at the right), eventually rounding off at the maturing face to form single, large cisterna-sized vesicles, apparently without fragmentation into small vesicles (compare *Zea,* below). Similar large vesicles (V) appear close to the plasma membrane (PM), and the plasma membrane bulges over packages of fibrillar material (e.g. black arrows) that closely resemble the contents of the vesicles. It is reasonable to conclude that this single micrograph shows many stages in the manufacture, packaging, movement, and delivery of the fibrillar material to its final destination. There it becomes more dispersed, probably as a result of hydration (asterisk). However the sample was chemically-fixed, and it is possible that the cisternae and large vesicles have suffered distortion by hydration of the contents during specimen preparation. Other noteworthy features include: (i) There is no obvious positional relationship between ER and the *cis*-face of the stacks, and no identifiable transitional vesicles (compare with Plate 10a and b). (ii) The *cis*-Golgi cisternae are composed of branched tubules (T), visible here in a stack that lies with its *cis*-face in the plane of the section. The inset shows the appearance of the same *cis*-face in the adjacent section, with extensions of the tubular system. (iii) Coated vesicles occur near the Golgi bodies, and two others are seen at the plasma membrane (circled), probably being formed at it (see also Plate 15). x30,000.

Plate 12b The ultra-thin section shown here (the periphery of a root cap cell of *Zea mays*) was oxidised with periodic acid to produce aldehyde groups in polysaccharides. Thiocarbohydrazide was then coupled to the aldehyde groups. This then allowed the electron-dense reagent silver proteinate to be used to label the polysaccharides by attahing it to the thiocarbohydrazide residues. Polysaccharides occur, as expected, in the cell wall (CW) and conspicuously in Golgi cisternae, thus confirming the suggestion that the latter are sites of polysaccharide biosynthesis. The inset shows that the polysaccharide does not have the same appearance in the cisternae and nearby vesicles as in the wall (bottom right hand corner of inset), thus indicating that further maturation and hydration takes place between the two locations. Vesicle formation at the *trans*-faces differs from that in *Phleum*, where the secretory packages seem to be whole cisternae. The stack under the inset shows stained polysaccharide in the successive cisternae in *Zea,* and then, at the *trans*-face, its packaging into peripheral vesicles (solid arrows), leaving compacted membrane fragments (open arrow) derived from the central region of the cisterna.

As in *Phleum,* there are no clear positional associations between ER and Golgi bodies and in neither case can medial-cisternae be distinguished. Nevertheless experiments in which radioactive substrate is tracked through the membrane systems indicate that transfer from ER to Golgi does occur. Label is incorporated rapidly into intermediate(s) in the ER and transferred after a few minutes to the Golgi apparatus. It reaches the cell wall compartment about ten minutes later, having been incorporated into high molecular weight (2.10^6) acidic and neutral polysaccharides. Maize root tip slime is distinctive in containing the sugar fucose, and tracer experiments show that this is added in the Golgi apparatus. x15,000, inset x36,000. Kindly provided by M. Rougier; (b) reproduced by permission from *Journal de Microscopie,* **10**, 67-82, 1972.

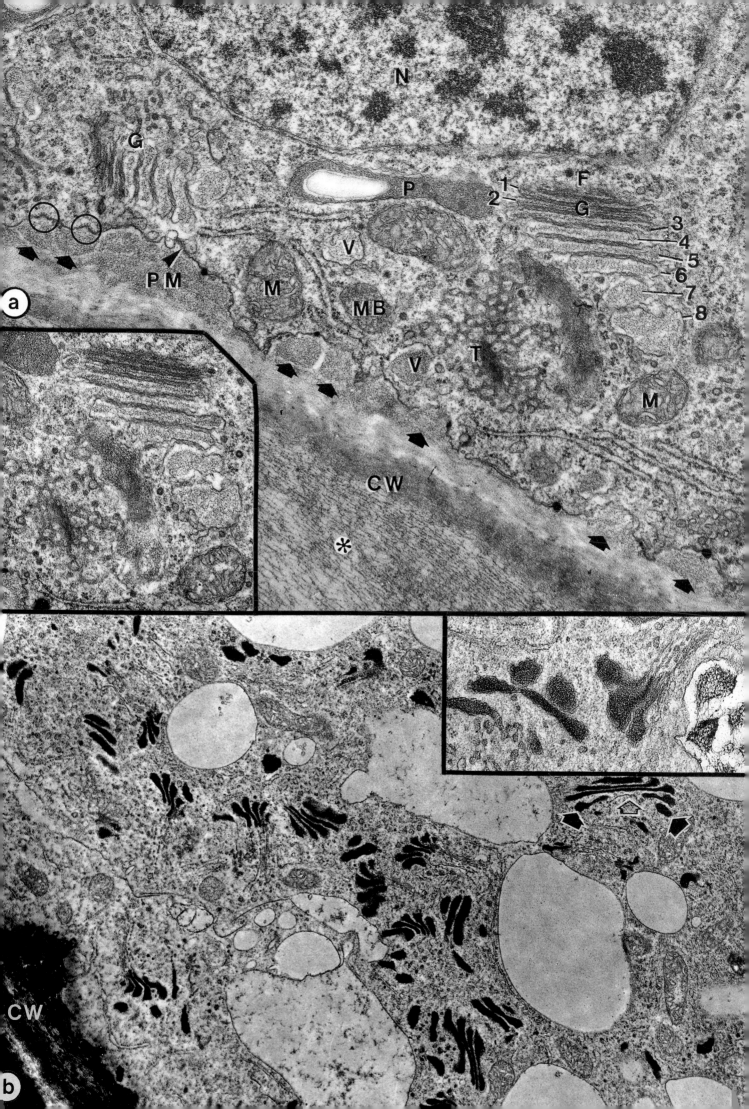

13   Plant Golgi Stacks: Compartments and Assembly

Introduction: Plates 9,11 and 12 illustrate gradients in morphology of the successive cisternae of plant Golgi stacks. The restricted distribution of particular enzymes or chemical groups is further evidence of heterogeneity of cisternae and compartmentation within each stack. Examples are presented here to reinforce the concept that Golgi stacks are analogous to factory assembly lines for stepwise fabrication of complex products.

Plate 13a Preservation of this Golgi stack in a tobacco (*Nicotiana*) root cap cell was optimised by rapid freezing followed by freeze substitution. The stack shows a distinct polarity from *cis*-face (C), where the cisternae are lightly stained with a wide lumen, to medial cisternae (M), which have densely stained luminal contents, to the *trans*-face (T), where the cisternae have a very narrow lumen and secretory vesicles form at their periphery, and finally the *trans*-Golgi network (TGN), which is likely to be a derivative of the *trans*-most cisterna, sloughed off the stack and carrying many coated and non-coated vesicles. The dark staining dots and short lines lying between the cisternae in the medial and *trans* regions are sections of sets of parallel intercisternal rods, believed to help cement adjacent cisternae to one another, and therefore important for maintaining the integrity of the stack. Forming vesicles acquire dense, non-clathrin coats (open arrows; see Plate 15 for further details). x80,000. Kindly provided by A. Staehelin, T. Giddings, J. Kiss and F. Sack, reproduced by permission from *Protoplasma* **157**, 75-91, 1990).

Plate 13b Assembly of complex polysaccharides occurs by sequential addition of monomer sugars by *transferase* enzymes. In general, transferases involved in early steps are found in the ER and the *cis*-cisternae, while transferases involved in completing the sugar chains are found in the *trans*-cisternae. Sycamore (*Acer pseudoplatanus*) cells grown in suspension culture were rapidly frozen at high-pressure and embedded in a resin designed to preserve the antigenic properties of their macromolecules. Sections were labelled with two different antibodies to see where two transferase steps are located in the Golgi stacks. One antibody, linked to large colloidal gold particles (large arrows), detects xyloglucan groups and the other, linked to small colloidal gold particles (small arrows), is specific for residues of the sugar fucose that become linked to the first N-acetylglucosamine residue of N-linked oligosaccharides in glycoproteins. The finding that both labels lie over the *trans*-cisternae and the *trans*-Golgi network indicates that enzymes for completion of xyloglucan synthesis and addition of terminal fucose to N-glycans are concentrated in these compartments. In other localisations, addition of fucose to glycoproteins occurs in medial cisternae, while stages of synthesis of the major pectin of the cell wall matrix occur throughout the stack. x11,500. Kindly provided by A. Staehelin.

Plate 13c,d Compartmentation in Golgi stacks is also revealed by treating samples with a zinc-iodide-osmium complex during fixation. In (c) (*Zea mays* root cap cell) reducing activity in the ER and the *cis*-cisterna of the Golgi stack forms an electron opaque precipitate. Medial cisternae contain less of the reaction product and *trans*-cisternae none at all. The stain is completely absent from vesicles (V), swollen with polysaccharide slime produced by the stack (see also Plate 12). x66,000

In (d) a small, stained *cis*-cisterna lies on the transitional zone side of a more heavily stained larger *cis*-cisterna. The medial-and *trans*-cisternae are not stained. Note the intercisternal rods between the cisternae. Cisternae of cortical ER (Plate 8) lie adjacent to the plasma membrane. Although ER strands pass through plasmodesmata (PD) as *desmotubules* (see Plate 45), the stain is restricted to a fine thread within the plasmodesmata, evidence that the lumen of their desmotubules is very narrow. x70,000; (c,d) kindly provided by C. Hawes, (c) reproduced by permission from *Bot. Gaz.* **143**, 135-145, 1982.

Plate 13e The successive cisternae in a stack are sequentially-functional and structurally-distinctive. It is not known how they maintain their structural association in the face of the very rapid entry of membrane and substrate from the ER, the shuttling of vesicles between successive cisternae, and the equally rapid exit of vesicles and products at the *trans*-face (for an example of data, see Plate 12). Treatment of plant cells with brefeldin A, a fungal metabolite that interrupts the transport of vesicles from ER to *cis*-cisternae, demonstrates how rapidly the stacks can be broken down and reassembled.

These three fluorescence micrographs of whole cells from maize root tips show the distribution of a fluorescein-labeled antibody, JIM-84, that binds to antigens in plant Golgi stacks. On the left, a control cell shows some of the several hundred Golgi stacks that are a normal feature of these cells. The dark area in the centre is the unstained cell nucleus. The middle frame shows a further group of cells that had been treated with brefeldin-A for one hour before staining. This disrupts transport processes from the ER to the Golgi apparatus and consequently leads to breakdown of Golgi stacks. Golgi membranes survive, but become associated in a relatively few very large clumps in the treated cells. Another set of treated cells (right hand frame) was allowed to recover from brefeldin-A treatment for two hours before staining with JIM 84. A normally distributed set of Golgi bodies has been restored in the post-treatment period. This experiment demonstrates that drastic alterations to the stack of cisternae can be induced and reversed in a relatively short timespan. x1,000. Kindly provided by B. Satiat-Jeunemaitre and C. Hawes, reproduced by permission from *The Plant Cell* **6**, 463-467, 1994.

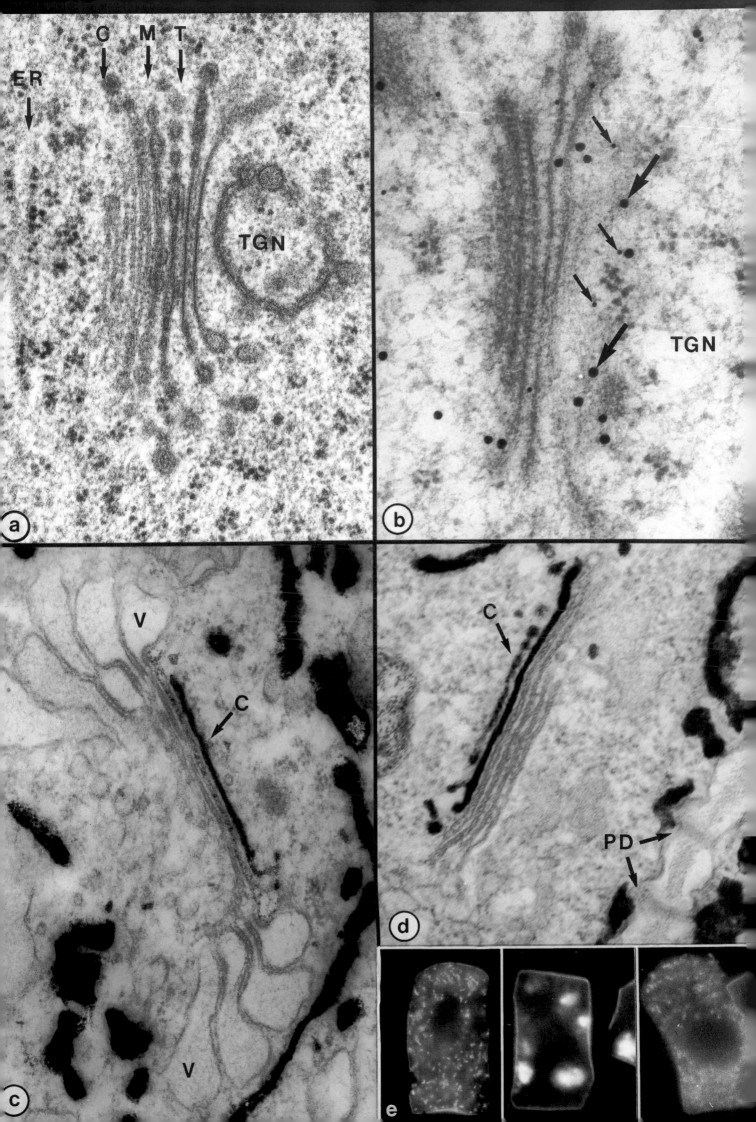

14 Protein Targeting in the Endomembrane System

Introduction: These micrographs illustrate aspects of *targeting*: how products made in one part of the cell are directed to other locations. One category of targeting concerns proteins that move through the ER - Golgi system. This category is considered here, using targetting of *vicilin,* a storage protein in legume seeds, as the example. Another category concerns proteins that are made on free ribosomes without entering the endomembrane system. An example of a protein that is targeted to mitochondria is described in Plate 19.

Plate 14a This cell in a cotyledon of a developing *Lathyrus* (sweet pea) seed has a central nucleus (N), with chromatin and a large nucleolus, surrounded by extensive ER (open arrows), starch grains (S) and protein bodies (PB), which are vacuoles filled with storage proteins, including vicilin. In the following pictures vicilin is localised by immuno-gold labelling. x660.

Plate 14b The source of the vicilin is the ER, many cisternae of which slant from lower left to upper right in this electron micrograph. Occasional gold particles show that vicilin is present either in the lumen or at the membrane of the ER. The region marked with an asterisk shows spiral polyribosomes on oblique surface views of the ER membrane. They are known to extrude growing vicilin molecules through the membrane into the lumen of the ER. Numerous gold particles over the vacuole (V) confirm that this is the site of ultimate accumulation of vicilin. To get from the ER to the vacuole the vicilin is packaged in small vesicles, processed through the Golgi apparatus, where it is modified and concentrated, and passed on, in other vesicles, to the vacuole. The inset shows part of a Golgi stack and its associated vesicles. Gold particles again reveal the presence of vicilin. Cells of developing *Pisum* (pea) seeds. (b) x24,000, inset x20,600.

The route from ribosomes to ER to Golgi to storage vacuole involves sequential guidance mechanisms. Vicilin, like other proteins that enter the ER, is targeted to the lumen by a *signal sequence* of amino acids. This sequence is the first part of the protein (at the carboxy terminus) to be assembled by the ribosomes as they translate the genetic code for vicilin (see Plate 7c). The signal sequence is clipped off in the lumen after it has played its part. Receptors in the ER then collect the vicilin molecules. They are packaged in membrane-bound vesicles equipped with surface molecules that in turn can be recognized by further receptors at the *cis*-face of the Golgi apparatus.

The chain of receptors and recognition events guides the vesicles and their enclosed vicilin to the Golgi stack, where vicilin is glycosylated, and then onwards to the vacuole. (c) explores how proteins that remain in the ER are sorted from those that move on to the Golgi apparatus and (d) explores targeting from the Golgi apparatus to the vacuole. While the *leading* signal sequence of amino acids determines whether a new protein enters the ER or remains in the cytoplasm, another signal at the *trailing* end (the amino terminus) of the protein determines whether the protein is to be retained in the ER or packaged and transported onwards. This *retention signal sequence* consists of the four amino acids: lysine (or histidine)-aspartic acid-glutamic acid-leucine. Receptors that encounter this sequence ensure that the proteins bearing it stay in (or are returned to) the ER. Antibodies to the sequence are useful in another context for staining the ER (Plate 36).

Plate 14c Vicilin does not normally carry the ER retention signal and hence it is normally moved on to the Golgi apparatus. This electron micrograph illustrates the outcome of an experiment in which the gene for vicilin was taken from pea plants. The genetic code for the four amino acids of the ER retention signal was added to it. The modified gene was adjusted further by adding a control region that ensured that it would be activated in leaf cells. It was then inserted into tobacco plants. The micrograph shows part of a leaf cell from one genetically-transformed tobacco plant. The immuno-gold labelling over a dense globule in the ER shows (i) that tobacco will make pea vicilin, given the necessary genetic information, (ii) that the synthesis can be switched on in leaves (and not just cotyledons, where it normally occurs), and (iii) that the modified vicilin is retained in the ER because of the presence of the retention signal. The arrowheads show the outline of the ER membrane and its continuity with the nuclear envelope. No accumulation of vicilin occurred in the vacuole or cytosol or chloroplast (C). Three gold particles over a mitochondrion (M) are contamination. x70,000.

Plate 14d Vesicles leaving the Golgi apparatus can move to the plasma membrane or to the vacuole. Just as export from the ER is a *default* pathway that operates when no retention signal is present, so the path from the *trans*-Golgi region to the plasma membrane may operate unless there are overriding signals to guide vesicles to the vacuoles. Several factors influence this traffic control, among them the ionic contents of the various compartments. This micrograph is from an experiment in which pea cotyledon cells were treated with a drug, monensin, that allows sodium ions to permeate through membranes. The gold particles show that vicilin transport has been diverted from its normal course. Much vicilin has still reached the vacuole (V, right) but much has taken the default route to the plasma membrane, where it has been exocytosed and deposited between the wall and the plasma membrane. x20,600.

(b and inset) and (d) kindly provided by S. Craig; (b inset) reproduced by permission from *Protoplasma* **122**, 35-44, 1984, and (d) from *Protoplasma* **122,** 91-97, 1984; (c) kindly provided by T.J. Higgins, reproduced by permission from *The Plant Journal* **2**, 181-192, 1992.

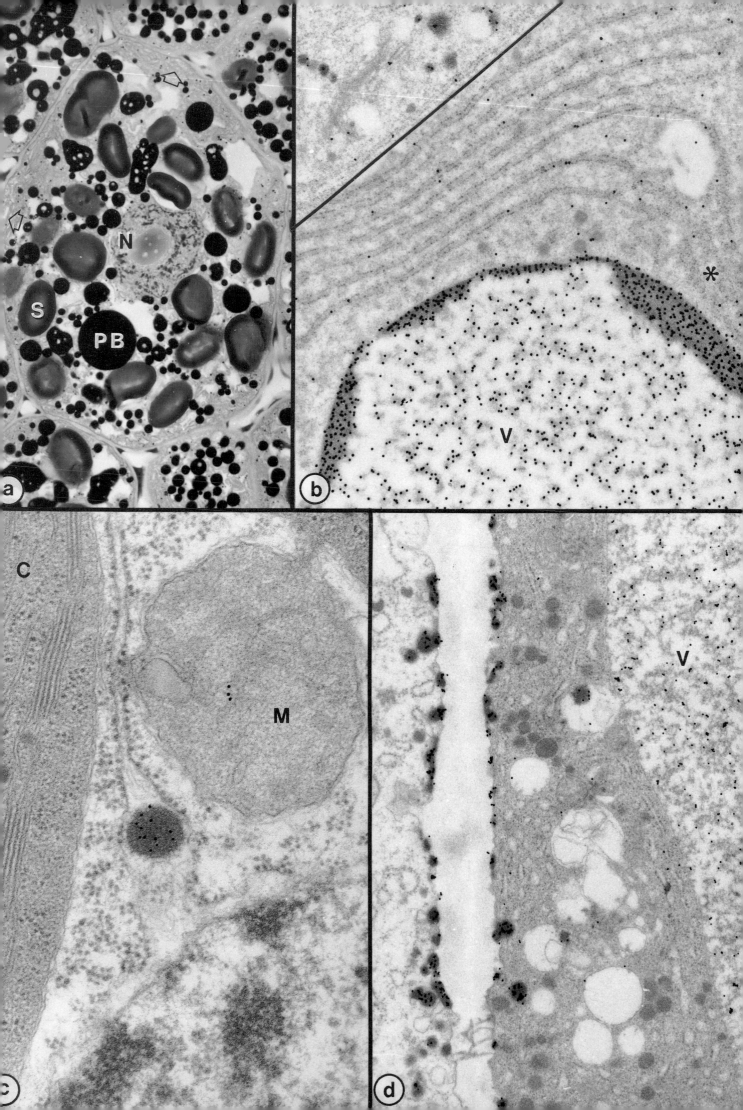

15 Coated Vesicles

Introduction: Subcellular cooperation between the various components of cells, each one with its specialised functions and unable to survive on its own, is essential for cellular life. Plate 14 gives one example of cooperation by means of signals and receptors that target proteins from their site of synthesis to particular destinations. Plate 19 deals with another aspect of protein targeting. Here we look at one aspect of a different kind of targeting, delivery of products to defined destinations after they have been packaged in vesicles. In these cases it is not sufficient to mark the products with signals, because they are hidden within the vesicles and hence cannot be accessed directly. Signals that specify the destination of the products must also be placed on the cytoplasmic face of the vesicles, where they can be perceived by a targeting mechanism. Microscopists see part of this targeting mechanism in the form of fuzzy coats on certain categories of vesicle - known as *coated vesicles*.

The fuzzy coats are proteinaceous, and of several sorts, of which the best studied is the protein *clathrin*. They may serve at least three functions. One is to collect molecules that are to be packaged in the vesicle, another is to generate the mechanical force that is required to deform a flat membrane and then pinch it off into a vesicle, and the third is to provide information for targeting, in conjunction with other membrane and soluble co-factors. Coated vesicles are shown in Plates 10, 12 and 13, during formation of vesicles at the ER and nuclear envelope (targeted to Golgi *cis*-cisternae), at the margins of successive Golgi cisternae, in the *trans*-Golgi network, and at the plasma membrane. This plate explores the class of clathrin-coated vesicles. They are found particularly at the plasma membrane. Vesicles with non-clathrin coats are commoner around the Golgi apparatus: the drug brefeldin A, whose dramatic effects on the Golgi apparatus are illustrated in Plate 13e, may be an inhibitor of their formation.

Plate 15a Clathrin protein forms polygonal lattice structures on the cytoplasmic face of the plasma membrane. In thin section the sectioned lattice appears as a series of short spikes. Two stages of invagination of these coated patches to form coated pits are depicted here. Later the pit pinches off as a free vesicle, still invested by the coat. Substances bound to receptors on the external face of the plasma membrane of the pit will be internalised when the vesicle forms, as will any substance in the fluid phase lying against the plasma membrane. Experiments have shown that small electron opaque particles, such as ferritin protein molecules and colloidal gold particles, or large molecules tagged with a fluorescent marker, can become trapped in these pits and internalised in the resultant coated vesicles. However it is also possible that this form of vesiculation is used to retrieve membrane from the cell surface, without necessarily packaging any contents. x90,000.

Plate 15b Cells attached to electron microscope grids can be chemically fixed, critical-point dried, and the bulk of the cell torn away to leave the plasma membrane and associated structures on the grid. Whole coated pits are seen face on (arrows) in this micrograph from such a preparation. In relation to the very high rates of membrane retrieval that are needed in some situations (see introduction to Plate 12), note the high density of these structures on the membrane surface (16 lie on the $2\mu m^2$ of membrane shown here). Some pits had just pinched off to form vesicles at the moment of fixation. They form more bulky objects for microscopy (solid arrowheads), but the projecting lattice on their surfaces is still discernible. The long straight structures running across the figure are microtubules. Carrot suspension culture cell, x70,000.

Plate 15c and d Unfixed, rapidly frozen cells can be fractured and the exposed surface etched and replicated. Rotary shadowing of the replica reveals images of clathrin coats with a good level of resolution. (c) shows a pit on the plasma membrane surface. The clathrin lattice forms an array of hexagons and pentagons. In this example there appear to be connections from the pit running across to an adjacent microtubule. Coated vesicles are also found deep in the cytoplasm, as in (d) where they are adjacent to the edge of a Golgi cisterna. One is seen in surface view (arrows), giving a good indication of the way the lattice is curved over the surface of the vesicle to form a cage, the other has been fractured through the middle (arrowheads), revealing the vesicle inside and showing the thickness of the lattice coat with radiating crevices between the clathrin protein molecules. c) x180,000; (d) x170,000.

Plate 15e Coated vesicles can be purified to yield preparations suitable for analysis of their proteins. This micrograph shows a section through an isolated coated vesicle. The clathrin proteins, now fixed and stained, again appear as projections radiating from the membrane surface. Comparison with the fractured vesicle in (d) shows that the chemically-fixed image is misleading, the "holes" in the lattice are in fact small compared with the lattice network of protein. x340,000.

Plate 15f and g The shape of isolated clathrin molecules (seen here after rotary shadowing) indicates how they associate with one another to make up the polygonal coat. They form three armed complexes termed *triskelions*. Each arm, 60nm long, radiates from a central vertex, with a terminal globule and a kink about halfway along. When triskelions associate, the ends of the arms overlap to form the edges of the lattice. Lattice corners are formed by the angle at the vertices and by the kinks half way along the arms. (f) x210,000; (g) x360,000.

All micrographs kindly provided by C. Hawes; (a), (d), (e), (f), and (g) reproduced by permission from *J. Cell Sci.* **88**, 35-45, 1987.

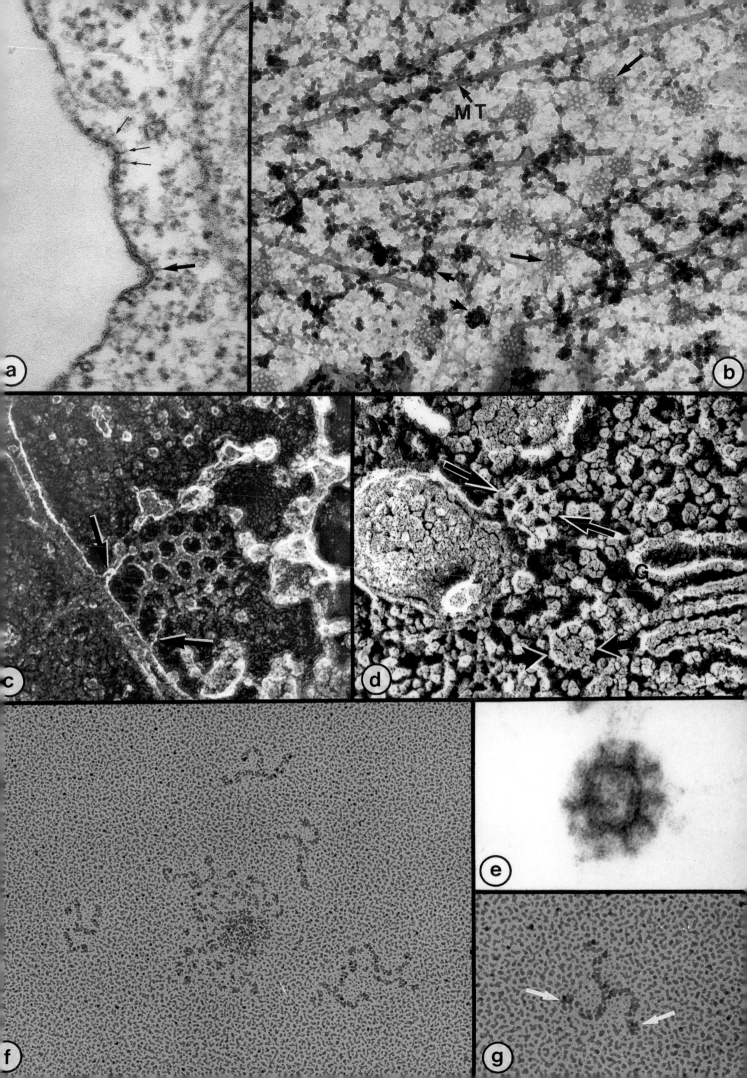

16 Vacuoles, Plasmolysis and Links between the Cytoplasm and the Cell Wall

Introduction: Vacuoles function in storage, digestion, osmoregulation and space-filling. Many types of substance are stored for short- or long-terms, e.g. seed protein reserves (Plate 14), coloured pigments (anthocyanins) in petals and leaves, organic acids that serve as metabolic pools and in pH regulation, and "defence" chemicals ranging from secondary plant products of innumerable sorts to proteins that inhibit digestion in herbivores. Some of the plant's own digestive enzymes are sequestered in vacuoles, in which cases the vacuoles function like lysosomes in animal cells, digesting compounds and even large cell components that are exocytosed into vacuoles through the tonoplast. Osmotically active substances are loaded into vacuoles by tonoplast transport proteins. Consequent inflow of water generates *turgor pressure,* stretching the cell walls (making the cells *turgid*) and thereby giving soft plant tissues mechanical rigidity. In most mature tissues vacuoles are the largest cell compartments, often occupying more than 90% of the cell volume. Their development drives cell expansion, and by spreading a little cytoplasm over a large surface area, they enable plants to attain large sizes with less expediture of energy than if the same volume had to be filled with cytoplasm. Vacuolar solutes and space-filling thus have a major role in determining the architecture and size of plants.

Removal of water from cells by *plasmolysis*, which occurs when they are bathed in media of low water potential (high concentration of osmotically-active solutes), causes loss of turgor. The protoplast shrinks and eventually the plasma membrane withdraws from the cell wall, at which time hidden links between the wall, the plasma membrane and cytoplasmic elements are revealed. These linkages have been known for a long time, but their nature and functions are now being studied anew.

Plate 16a   Tonoplast: In this freeze-etched cell, the fracture plane has passed over one vacuole (left) and under another (right), passing along the mid-line of the tonoplast membranes. The resulting convex and concave surfaces represent the vacuolar (V) and cytoplasmic faces (C) of the tonoplast. The asymmetrically distributed intra-membrane particles possibly include membrane ATPases that drive solute transport across the tonoplast. x60,000, Kindly provided by B. Fineran, reproduced by permission from *J. Ultrast. Res.* **33**, 574-586, 1970.

Plate 16b   Plasmolysis: Live onion scale-leaf epidermis was stained with $DIOC_6$ to make cell membranes fluorescent (also used in c,d; for details of this reagent see Plate 36q,r). A group of cells was photographed before (top) and after (below) plasmolysis in 0.75M mannitol. The cell walls are obscured by brightly fluorescent cortical cytoplasm before plasmolysis, but are visible after plasmolysis against the dim fluorescence of the staining fluid that fills the space between the walls and the shrunken protoplasts. x120.

Plate 16c Hechtian strands of stretched plasma membrane extend from the plasmolysed protoplast to the cell wall (K. Hecht, in 1912, was the first to study them in detail). This confocal micrograph shows strands in a 16μm-deep zone. Some strands extend from plasmo-desmata (arrows), where the membrane passes through the cell wall, but many attach to other wall sites. In 1931 another early investigator, J. Plowe, pushed plasmolysed protoplasts along inside their cell walls, observing that *existing* strands stretched and broke, and that *new* strands could form. In today's terminology we would say that new strands arise when free *binding sites* on the plasma membrane contact and bind their *ligands* in the cell wall.

Plate 16d   These two confocal optical slices of a group of plasmolysed cells include the upper surface of the cells (right), and an adjacent deeper layer (left). Hechtian strands can branch (arrow). At their outer extremity they may link either to wall sites (not distinguished morphologically), to plasmodesmata, or to portions of cortical endoplasmic reticulum (CER) (arrowheads) that remain attached to the inner surface of the cell wall, still in their original polygonal networks (see Plate 8). Images of this type indicate that there are molecules which attach the plasma membrane to components of the cell wall, and which can also traverse the plasma membrane to attach to internal elements in the cortical cytoplasm. CER (shown here) and actin (see (e)) are two such elements. The CER can be so strongly bonded through the plasma membrane to the wall that it can be held in place during plasmolysis. The plasma membrane wraps over it as it withdraws, and is pulled out into a Hechtian strand. (c,d) x1,250.

Plate 16e   This epidermal cell from a young *Tradescantia* leaf was microinjected with rhodamine-labelled phalloidin to label its actin cytoskeleton (for details see Plate 35) while it was still alive and turgid (left). It was then plasmolysed gently (right). The plasma membrane withdrew from the cell wall in several hemispherical regions, compressing the system of actin microfilaments. Arrows point to fine strands of actin stretching from the plasmolysed protoplast towards the cell wall. It seems that as well as trans-plasma membrane links from the wall to the CER, there are links from the wall through the plasma membrane to cytoplasmic actin microfilaments. x1,000, kindly provided by A. Cleary.

The molecular links inferred from (d,e) are analogous to *integrins*, trans-plasma membrane proteins in animal cells which transmit signals from the extracellular matrix to the cytoplasm. They could have important functions in plants, particularly in transmitting mechanical stimuli, e.g. in responses to touch (thigmotropisms), in gravitational responses, and to enable cells to perceive mechanical strain in the wall, perhaps participating in turgor sensing and feedback mechanisms which influence cell growth, e.g. *via* rearrangements of cortical microtubules (see Plate 33).

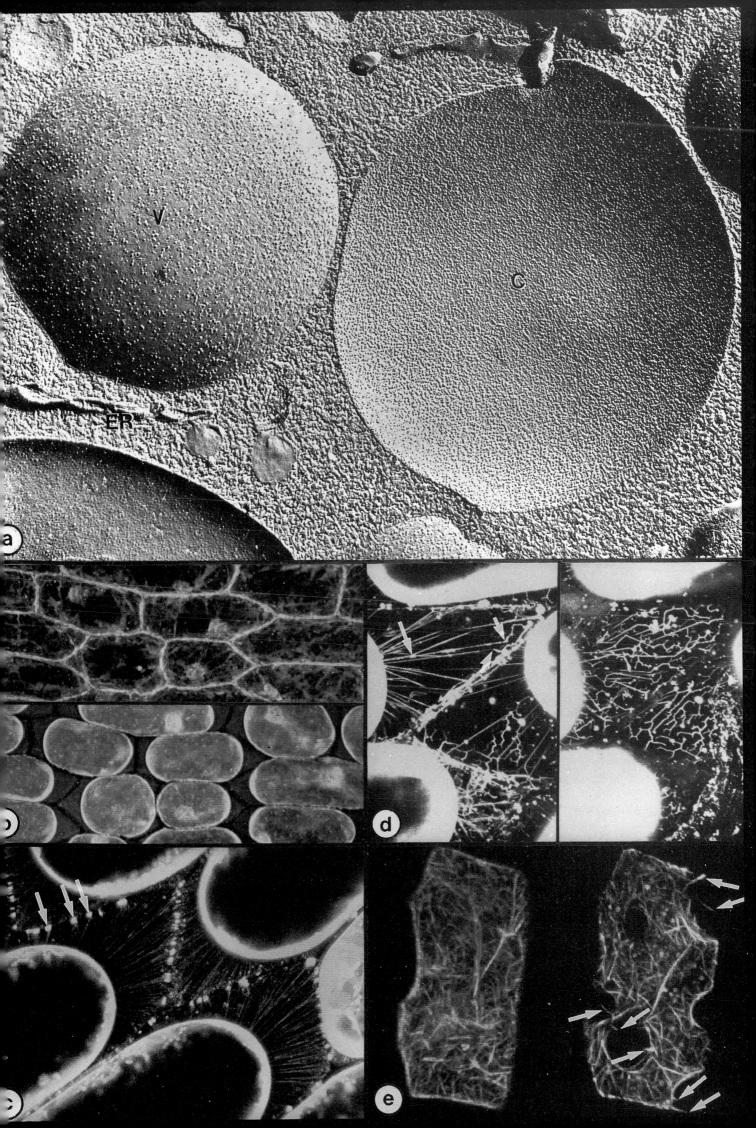

17 Membranes of Mitochondria

Introduction: Mitochondria consist of two envelope membranes bounding an inner space, the *mitochondrial matrix*. While the outer membrane forms a smooth curved surface in contact with the cytoplasm, the inner membrane is thrown into a series of folds, the *cristae*, which project into the matrix. Mitochondria are cylindrical to ellipsoid or ovoid, sometimes elongated and even branched. In living cells they are often observed to change shape, i.e. they are *pleiomorphic* (see next plate).

The matrix contains a nucleoplasm region containing DNA. The DNA and mitochondrial ribosomes, also in the matrix, both resemble their counterparts in prokaryotic bacteria (see Plate 19). Mitochondria can only increase in number by fission of the existing structures. Most plant cells contain several hundred mitochondria, but some unicellular algae contains only a single mitochondrion with a complex, reticulate shape (Plate 18).

Mitochondria are the sites of cellular respiration, involving the uptake of oxygen coupled to generation of the main immediate source of chemical energy in cells, ATP, and the formation of carbon dioxide. These two events result from different sets of biochemical reactions located in different parts of the mitochondrion.

Glycolytic processes within the cytoplasm break down sugars to smaller, pyruvate molecules that can pass across the mitochondrial envelope to the matrix, where they meet a series of enzymes that make up the *Krebs*, or *tricarboxylic acid, cycle*. Substrate molecules entering the matrix are decarboxylated to two-carbon acetyl units which are oxidised by Krebs cycle reactions to CO_2.

The oxidation steps transfer hydrogen from substrates to carrier molecules. In turn electron transport systems on, and within, the membranes of the cristae separate the hydrogen into electrons and protons. The protons are released into the space *between the membranes*, within the folds of the cristae, and the electrons are passed along a chain of carriers in the membrane to cytochrome proteins that terminate the chain. The cytochromes catalyse recombination of electrons with protons *derived from the matrix* and with oxygen to form water.

Release of protons into the inter-membrane space and subsequent consumption of protons from the matrix stores energy in the form of a *proton gradient* across the inner membrane. Protons flowing down this gradient, back into the matrix, drive ATP synthesis. The proton flow is channelled through large trans-membrane multi-protein complexes, ATP synthases, which traverse the cristae membranes and project into the matrix.

These functional arrangements are reflected in mitochondrial structure. The folding of the inner membrane provides a large surface area for housing the electron transport chain components and the ATP synthases. The folds also provide intimate contact with the matrix and Krebs cycle enzymes, and enlarge the capacity of the intermembrane space.

Plate 17a This group of mitochondria is lying in the cytoplasm of a parenchyma cell within the floral nectary of broad bean (*Vicia faba*). Although it is difficult to ascribe a precise function to this particular cell, the tissue as a whole secretes nectar, a solution containing mainly sucrose, in an energy-requiring process involving active transport across the plasma membrane. Most of the mitochondrial profiles seen here are circular, suggesting that the mitochondria approximate to spheres 0.75-1.5µm in diameter. Individual mitochondria are surrounded by an outer (O) and an inner (I) membrane, the latter being infolded (arrowheads) to form the cristae (C). In the matrix of many of the mitochondria there are electron-transparent areas, the nucleoids (N), which contain fine DNA-fibrils (F). Mitochondrial ribosomes (R) lie in the more densely stained regions of the matrix, but are inconspicuous. Mitochondrial granules (Gr) are more obvious, and probably consist of calcium phosphate. One of the mitochondria (asterisk) is linked to another (star) by a continuous outer membrane (O). This could represent a stage of mitochondrial division, but the same appearance could also be obtained in certain sections of a Y-shaped branched mitochondrion.

Other features are the plastid (P), and a face view of a Golgi cisterna (G) with its associated vesicles. There are numerous secretory vesicles, some quite large (V), and containing a fibrillar material. Many of the free cytoplasmic ribosomes are organized in polyribosomal helices (PR) (see also Plate 7c). x28,000.

Plate 17b This *transfer cell* is located alongside two xylem elements (X) in the cotyledonary node of a lettuce seedling (*Latuca sativa*). The cytoplasm adjacent to the wall-membrane apparatus (see also Plate 46) contains many mitochondria with densely packed cristae (C), which nearly completely obscure the nucleoids (N); otherwise they are similar to the mitochondria in (a). The high rate of respiration indicated by the number and conformation of these mitochondria may be connected with ATP requirements for pumping solutes across the plasma membrane of the transfer cell. x24,000.

Plate 17c These two mitochondria in a mesophyll cell of *Vicia faba* illustrate the inner (IM) and the outer (OM) membranes. The mitochondrial ribosomes (R) are smaller than their counterparts lying in the cytoplasm outside. Fibrils (F), probably of DNA, lie in the nucleoid regions. As in the other micrographs on this plate, the number of profiles of cristae (C) within the mitochondria far exceeds the number of visible connections (arrows) between the cristae and the inner mitochondrial envelope. The inference is that the cristae expand within the mitochondria from small connections to the inner envelope membrane - an arrangement that may be functionally efficient in reducing dissipation of the proton gradient away from ATP synthases located in the membranes of the cristae. x65,000.

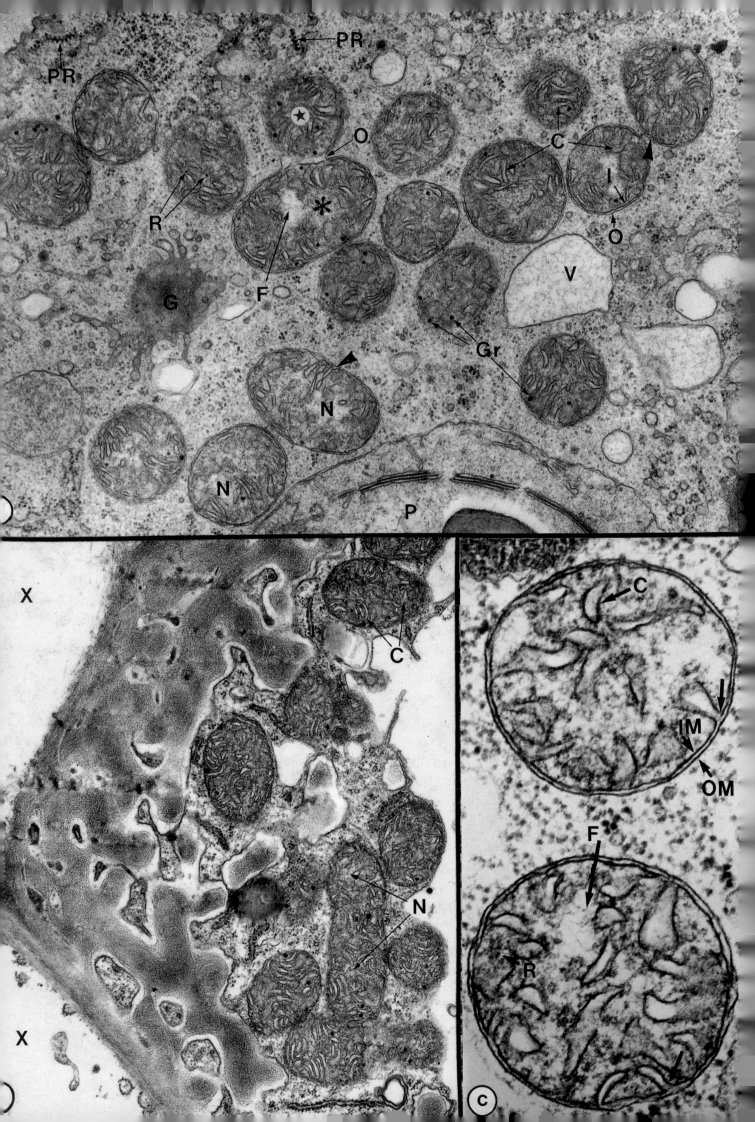

18 Variations in Mitochondrial Morphology

Introduction: Although the examples presented in Plate 17 are typical of plant mitochondria, other configurations occur. Some are displayed here.

Plate 18a,b These confocal fluorescence micrographs show mitochondria in tobacco (strain BY-2) cells that were grown in suspension culture. Rhodamine-123, a vital stain (i.e. a non-toxic one that can be applied to living cells), was used. It is accumulated by mitochondria that are maintaining a potential difference across their inner membrane by their metabolic activity. The staining is quite specific if dilute solutions are used. Plastids stain comparatively weakly. Higher concentrations can be used to reveal endoplasmic reticulum (Plate 8).

(a) is a low magnification optical slice, 5μm thick, showing mitochondria at the cell surface, in trans-vacuolar strands, and around the nucleus. The cell contained more than a thousand mitochondria. x700.

(b) this region of the cortical cytoplasm of a similar cell, at higher magnification, shows a variety of mitochondrial shapes and dimensions. The smallest and commonest are ellipsoids about 0.5μm by 1-2μm. The narrow dimension is quite constant, but they can elongate markedly (arrow). Long mitochondria sometimes show constrictions (arrowheads) where fission is about to occur. x1500

Plate 18c: These unusual mitochondria were found in some of the cells in the root tip of white lupin (*Lupinus albus*). Elucidation of the three-dimensional config-uration is impossible from a single micrograph. One mitochondrial profile (M-1) forms an irregular but continuous ring, with another (M-2) lying quite close to it. M-1 and M-2 are probably parts of a single extensive plate-shaped mitochondrion that undulates up and down through the plane of the section. The inset demonstrates that arms of the mitochondrion are connected (asterisk). If all parts of M-1 and M-2 are connected in this manner, the plate so formed must be at least 15μm x 4μm. The inset also shows numerous small cristae (C) and mitochondrial ribosomes lying in the mitochondrial matrix. Like chloroplast ribosomes (Plate 19), the latter are smaller than cytoplasmic ribosomes.

Comparable very large mitochondria have been found in female reproductive cells in some plants, and there it is estimated that the amount of DNA per mitochondrion is up to 4 megabase pairs, much more than the usual content. As noted in Plate 19, plant mitochondria typically possess about 200 kilobase pairs of DNA, which is more than most animal mitochondria. Those in meristematic cells have been reported to contain multiple copies of the basic DNA quota. Fission and enlargement without commensurate DNA synthesis as the cells divide and start to elongate gives rise to longer mitochondria with less DNA in single nucleoids. Chloroplasts, by contrast, become many times polyploid as they mature. Magnifications: x8,000, inset x18,000.

Plate 18d and e: The micrograph shown in (d) (x20,000) is from a sequence of 43 adjacent serial sections that together encompassed the whole of a young cell of the unicellular green alga *Chlorella fusca* var. *vacuolata*. The complete sequence was used to build a three-dimensional reconstruction of the mitochondrion and chloroplast shown in (d). The model is viewed as from the bottom of (c) looking upwards; (c) corresponds to the level in the model marked by the arrows X and Y. Components are labelled as follows: nucleus (N - dotted line in (e), nucleolus seen in (d)); vacuoles (V - containing electron-dense polyphosphate bodies); chloroplast (C - containing a pyrenoid (P) which is shown exposed in (e) by cutting away the layer of chloroplast that covers it); microbody (MB - lying in the cytoplasm, but in close proximity to the pyrenoid, a regular feature of these cells - see also Plate 23c,d). Although *Chlorella* is a non-motile alga (i.e. lacks flagella), it contains a pair of *centrioles* (CP) (or *basal bodies* - see Plate 32a for further details), which are probably relics of an ancestral motile form. The three layered (dark-light-dark) structure (W) in (d) is the remains of the wall of the parent cell. In this (but not all) species of *Chlorella*, the outer layer of the cell wall contains a substance that closely resembles *sporopollenin*, the resistant wall material of pollen (see Plate 56).

The main feature of the reconstruction is that the mitochondrion (M), seen as quite separate profiles (1-8) in (d), is a 3-dimensional continuum of loops and branches - a *mitochondrial reticulum*. Some branches lie close to the cell surface (2 and 3 in (d) = A and B in (e)). Another branch (7 in (d), C in (e)) penetrates a cytoplasmic channel that passes through the chloroplast, emerging at D. Three dimensional reconstructions have demonstrated that recticulate mitochondria are quite common in single-celled organisms. They do not normally feature in multicellular plants.

Plate 18f Ring-shaped mitochondrial profiles are commonly found in electron micrographs. They probably result when a shallow dish-shaped mitochondrion is sectioned across the dish so that only the rim is contained within the section. The two layers of the mitochondrial envelope are visible at both inner and outer faces of the ring (circles). In this example, from a developing cotyledon of *Phaseolus multiflorus*, the crista is unusually extensive and forms a continuous cisterna, at least in this part of the mitochondrion. x60,000.

Plate 18g These four mitochondria, from a developing pollen grain of *Avena fatua*, are very small and relatively undifferentiated. None is more than 0.5μm in diameter. Each contains a very few cristae (C) and mitochondrial ribosomes (R), within the inner (I) and outer (O) membranes. Both mitochondria and plastids can become reduced to a very rudimentary state during development of pollen. x55,000.

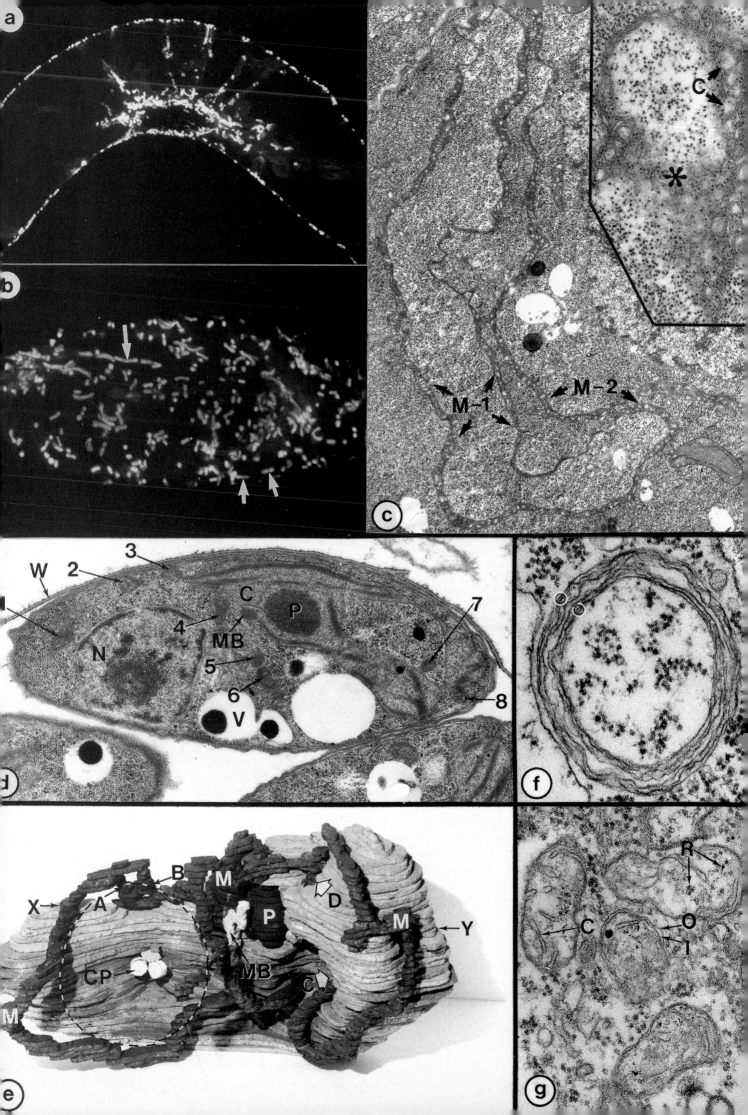

19   Partial Autonomy of Mitochondria and Plastids

Introduction: Mitochondria and plastids are believed to have arisen through evolutionary modification of prokaryotic organisms that entered into *endosymbiotic* relationships with eukaryotic host cells. Much of the genetic information of the introduced symbionts has by now been transferred to the host nuclei, so mitochondria and plastids are no longer able to synthesise all of their own nucleic acids and proteins. They remain *semi-autonomous*. This Plate illustrates their DNA and ribosomes, and how proteins are targeted to them.

Plate 19a   DNA (i): This region of a cross-section of an *Arabidopsis* root tip shows parts of seven cortical cells. The dye 4'6-diamidino-2-phenylindole (DAPI) was used to stain DNA, which appears as bright fluorescent areas. Four cell nuclei are present, each with a large central unstained nucleolus (Nu), surrounded by bright heterochromatin regions in less bright euchromatic nucleoplasm. The numerous spots of fluorescence in the cytoplasm represent DNA in mitochondria and proplastids. The two cannot be distinguished, but some of the smaller spots are mitochondrial DNA (of the order of 200 kilobase pairs, or 2.10^{-16}g each) and some of the larger ones are proplastid DNA. Mitochondria and plastids may contain more than one nucleoid area. The total quantity of DNA in a plant mitochondrion is nearly equal to that in a small bacterium. *Arabidopsis* is unusual in the small size of its haploid nuclear genome, 70-100 megabase pairs, i.e. $0.7-1.10^{-13}$g, so in cells where the plastid DNA has become amplified many-fold, as during chloroplast development in leaf cells (see (b)), the total amount of DNA in a cell's plastids and mitochondria adds up to nearly one third that in the nucleus, x3,000.

Plate 19b   DNA (ii): A nucleoid area traversed by fine DNA fibrils (F) in an ultra-thin section of a developing chloroplast. Many identical circles of DNA arise during chloroplast development, and are packed into this and other nucleoid spaces in the stroma compartment. Plastid ribosomes (R) are smaller than those in the cytoplasm outside the chloroplast envelope (CE). x75,000.

Plate 19c,d   Ribosomes (i): Plastid and mitochondrial ribosomes remain more like those of free-living prokaryotes than those of their eukaryotic host cells. They are smaller than eukaryotic cytoplasmic ribosomes (see above), as are their ribosomal RNA (rRNA) molecules. The distinction is illustrated here by using specific probes consisting of polyribonucleotides that distinguish between cytoplasmic and chloroplast rRNA molecules. In (c) antisense RNA to pea 18S rRNA (the largest of the rRNA molecules of *cytoplasmic* ribosomes) was linked to biotin and used to label sections of an alga, *Cryptomonas*. The biotin label was located by applying rabbit anti-biotin antibody, followed in turn by goat anti-rabbit antibody that had been labelled with electron-dense 15 nm gold particles (see diagram). The gold particles lie over the site of synthesis of the rRNA, the nucleolus, and over the sites

of this RNA in the cytoplasm, the ribosomes. In (d), the cells were labelled with a probe made from the 16S RNA of *chloroplast* ribosomes (the smaller counterpart of the 18S probe used in (c)). This probe labels the ribosomes in the chloroplasts only. Kindly provided by G. McFadden, I. Hooke and A. Clarke, reproduced by permission from *MICRO 90*, London, 683-688. x44,000

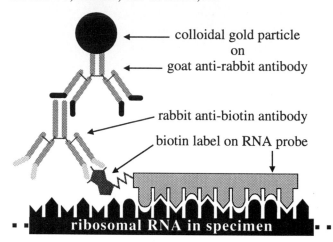

colloidal gold particle on
goat anti-rabbit antibody

rabbit anti-biotin antibody

biotin label on RNA probe

ribosomal RNA in specimen

Plate 19e   Ribosomes (ii): Ribosomes (R) in the matrix of a simple mitochondrion in an *Azolla* root tip cell. As with plastid DNA, mitochondrial DNA codes for just a limited number of proteins, assembled in the stroma using these ribosomes. Although few in number, the mitochondrial (and plastid) genes are essential. x68,000.

Plate 19f, g   Protein import: Most of the proteins of mitochondria and plastids are made in the host cytoplasm. *Signal sequences* of amino acids guide them to their destinations, like the sequences that guide other proteins into the endoplasmic reticulum (Plate 14). Outer membrane proteins enter that membrane directly. For others, receptors occur at localised import sites, where the two membranes of the mitochondrial envelope come into transient contact with each other (arrows). Some inner membrane proteins enter the matrix and then the membrane, others enter it from the inter-membrane space. Matrix proteins are guided through both membranes. Import sites are revealed here by supplying intact, isolated, swollen mitochondria of *Neurospora* with a protein of the ATP synthase complex, one of many proteins normally imported into mitochondria. The protein had been coupled to immunoglobulin molecules, which label the import sites by sticking in them and thus preventing completion of translocation. The immunoglobulins were marked with a gold-labelled reagent (protein-A-gold) and the sample embedded and sectioned. Arrowheads in (f) and in the higher magnification view in (g) indicate gold particles at import sites. Membrane contact sites (unlabelled) appear in Plate 17c. (f) x103,000; (g) x173,000. Kindly provided by M. Schwaiger, V. Herzog and W. Neupert, reproduced by permission from *J. Cell Biol.*, **105**, 243, 1987.

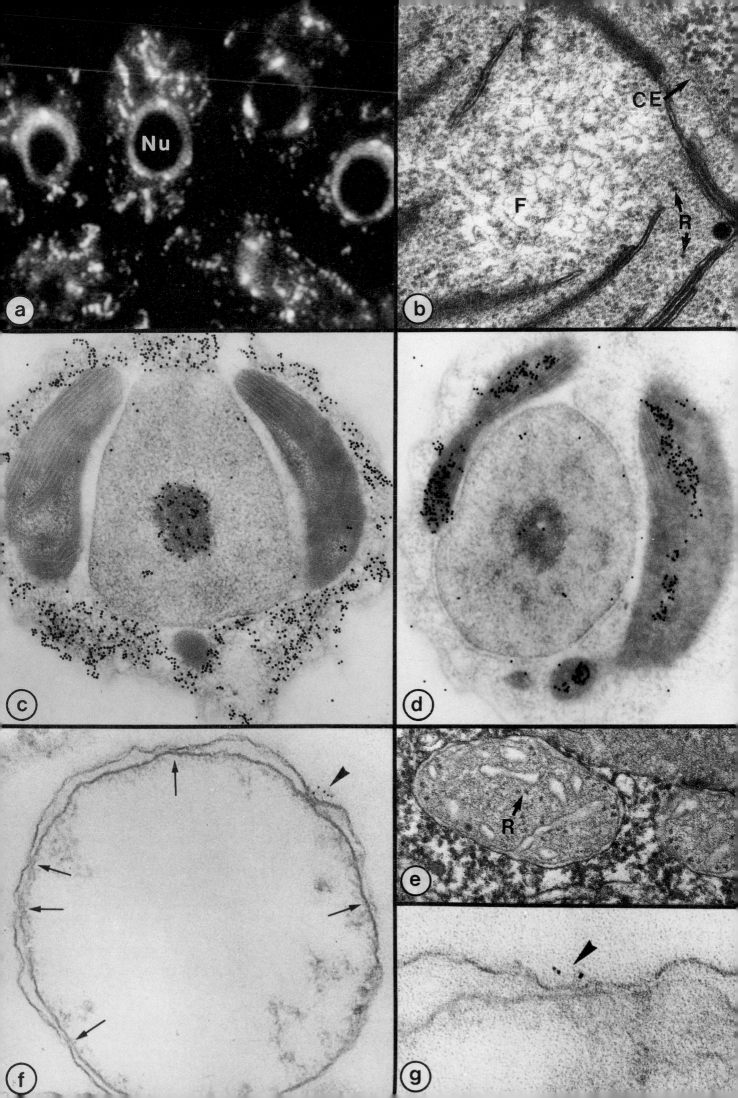

20 Plastids (I): Development of Proplastids to Etioplasts and Chloroplasts

Introduction: The presence of plastids is one of the features that distinguishes plant cells from animal cells. All plastids have a basic, general structure that is modified and differentiated as the plant cell develops to its mature state. Structurally we can recognise five or six different types of plastid, the simplest being the proplastid, a precursor form found in meristematic cells. Like other plastids, proplastids contain genetic information which, in concert with genes in the cell nucleus, guides development into one or other of the forms, i.e. *chloroplasts* in photosynthetic leaf tissue, starch-storing *amyloplasts* in storage tissue, yellow-orange pigmented *chromoplasts*, *etioplasts* in etiolated tissues, or rudimentary *leucoplasts* in non-green differentiated tissues, e.g. epidermis.

All plastids are bounded by an envelope consisting of outer and inner membranes, enclosing the inner compartment, the stroma. Internal membrane cisternae, known as stroma lamellae, may be present; they are especially well developed in chloroplasts, where they are known as *thylakoids* and carry the pigments and other components of the light-harvesting system of photosynthesis. The stroma contains plastid DNA in nucleoid areas, in the form of circles of DNA , and also plastid ribosomes, sometimes present as polyribosomes. Plastids can make some of their own proteins, but their autonomy is very restricted. Most plastid proteins are imported from the cytoplasm, guided to their destination by signal sequences of amino acids comparable to those involved in targeting proteins to mitochondria and sub-mitochondrial compartments (see Plate 19).

Plate 20a Plastids with simple internal structures, but with the capacity to develop into more complex types, are described as proplastids. This is the characteristic form of plastid in meristematic cells. They are illustrated here (P) in a cell at the outer surface of a stem apex, and in part of an underlying cell (bottom left of micrograph). They are similar in both cells, despite the fact that maturation and differentiation of the two will produce very different plastids - rudimentary non-pigmented leucoplasts in mature epidermis, and well developed chloroplasts in photosynthetic tissues. A nucleus (N), with nucleolus, a large vacuole (V) with flocculent contents, several smaller vacuoles, and the thin cuticle (C) are also seen. Oat (*Avena sativa*) stem apex, x5,300.

Plate 20b The proplastids of meristematic root tip cells are similar to those in cells at the stem apex (20a). Root tip proplastids, however, do not normally develop into chloroplasts. The one shown here displays the two concentric membranes of the envelope (visible, e.g., within the circles), with occasional invaginations of the inner membrane (arrows, left hand side). A small starch grain (S) is present in the section, partly ensheathed by internal membranes (thylakoids). The internal membrane system is sparse and not organized. A few particles which may be plastid ribosomes (smaller than cytoplasmic ribosomes) are seen (e.g. within square) and nucleoid areas (large arrows), with fine fibrils of DNA, are also present in the stroma. *Vicia faba* root tip, x77,000.

Plate 20c This micrograph is representative of the appearance of leaf cells which are differentiating in the light from the meristematic condition shown in 20a to become photosynthetically-active mesophyll cells. The vacuoles (V) are large, and intercellular spaces (IS) are enlarging, generating contact surfaces across which CO_2 will be absorbed during photosynthesis. Much of the chromatin in the nucleus (N) is heterochromatic (dense areas). The plastids (P) have developed complex internal membrane systems, with small grana interconnected by stroma lamellae, and by now are young chloroplasts. Some starch grains are present in them (S). *Avena sativa*, young leaf of illuminated seedling, x8,000.

Plate 20d The same tissue as for 20c was used to obtain this micrograph, but the seedling was grown in darkness so that it became etiolated. Vacuoles (V), intercellular spaces (IS), and nucleus (N) are much as in the light-grown seedling, but the plastids have developed into young etioplasts (E). The main feature of etioplasts is the semi-crystalline lattice of membranous tubes known as the *prolamellar body* (open arrows). The example numbered 1 is of the type illustrated further in Plate 26h, and number 2 is of the type shown in Plate 25e and g. Number 3 is probably of the same type as Plate 26g. If the plant were to be illuminated, etioplasts would pass through developmental stages of the type shown in Plate 27 and become chloroplasts. *Avena sativa*, young leaf of etiolated seedling, x 8,000.

20d reproduced by permission from *Biochemistry of Chloroplasts,* **II**, 655-676 (Ed. T.W. Goodwin, Academic Press, 1967).

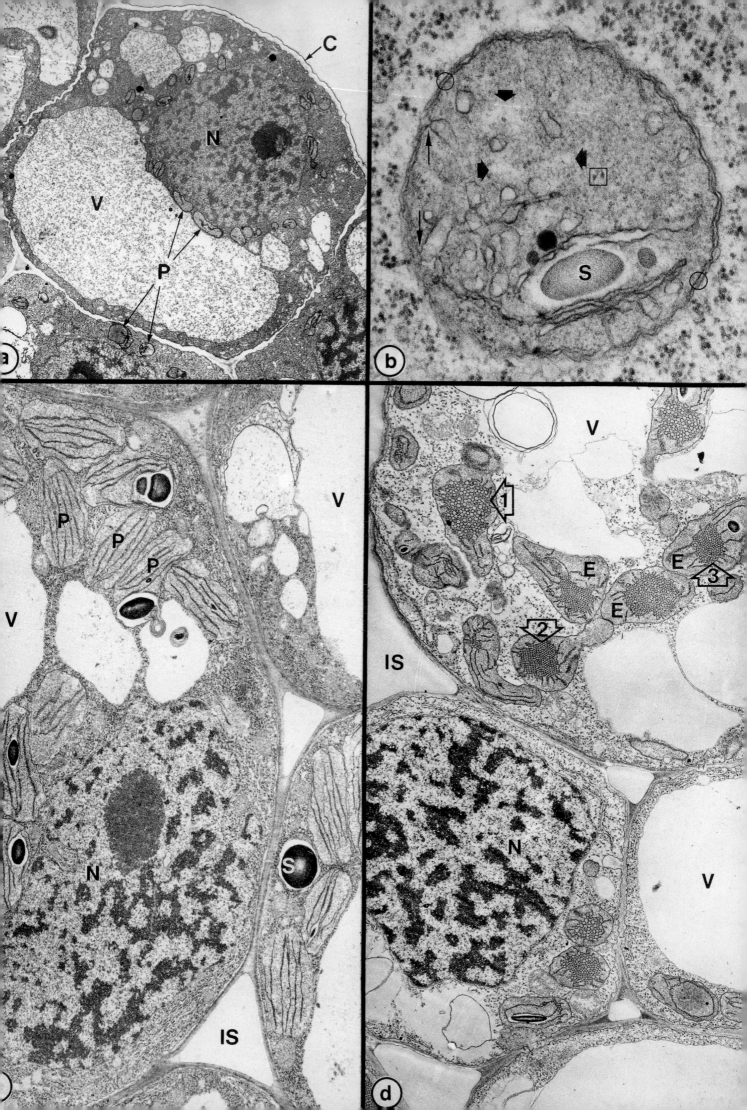

21 Plastids (2): Chloroplasts and Thylakoids

Introduction: The internal membranes of chloroplasts, known as thylakoids, form a very complex membrane system, regionally differentiated into lamellae that traverse the stroma and stacks of flattened discs, the *grana* (singular *granum*). Membranous branches called *frets* connect the stroma and grana lamellae. The arrangement creates four domains in chloroplasts, each with particular functions in photosynthesis: (i) the stroma, (ii) a labyrinthine intra-thylakoid lumen, (iii) an area of thylakoid membrane whose surface is exposed to the stroma, (iv) areas of appressed membranes where adjacent granum discs lie in contact with each another.

The *stroma* is the site of the biochemical carbon reduction reactions of photosynthesis that generate triose phosphate (a three-carbon sugar) from carbon dioxide. Triose phosphate is either converted to starch in the stroma (Plate 23) or transported to the cytoplasm and converted into sucrose (the major form in which organic nutrient is moved around plants in the phloem tissue (Plates 49,50)). The *thylakoids* are the site of the photochemical reactions of photosynthesis. They house light harvesting chlorophyll-protein complexes that funnel light energy to two different photoreaction centres. Light-driven transport of electrons along chains of donor and receptor molecules begins at these centres, generating the two products, $NADPH_2$ and ATP, that are required in the stroma for carbon reduction. The two photoreaction centres are differentially distributed between the two domains of the thylakoid surface (iii and iv above). The labyrinthine intra-thylakoid space is important because protons are moved into it from the stroma during the electron transport reactions. Then, much as in mitochondrial ATP synthesis, the proton gradient discharges back to the stroma over the extensive thylakoid-stroma contact surface, through ATP synthase enzymes that exploit the energy of the gradient to produce ATP.

Plate 21a A transverse section of a typical leaf with Calvin cycle photosynthesis is viewed here by light microscopy. The cells of the upper and lower epidermis do not contain well developed chloroplasts. These are visible, however, in cells of the spongy mesophyll (S) and the palisade layer of columnar cells (P). In the small vein on the left it can be seen that chloroplasts are not obvious in the bundle sheath cells (stars), unlike the bundle sheath in the "C4" plant in Plate 23a. *Hypochaeris radicata*, x165.

Plate 21b One palisade cell from the same material as in 21a is shown in a 1-2μm thick section that passed through the peripheral layer of chloroplasts in the cell. Numerous grana (the dark grains) are resolved within each chloroplast. The hemispherical shape of the chloroplasts can be inferred from the two types of profile present, the side views correspond to the plane shown in 21c, the circular top views to that in 21d. x1,100.

Plate 21c and d These electron micrographs of chloroplasts show the two membranes of the chloroplast envelope (E); side (S) and top (T) views of grana; densely staining ribosomes in the stroma (circled); starch grains (G), the fret membranes (F) that interconnect the grana (better seen in 21c than in 21d, where most of them are present in oblique or face view); and plastoglobuli (P). (c) *Avena ventricosa*, x33,000; reproduced by permission from *Can. J. Genet. Cytol.* **12**, 21-27, 1970; (d) *Zea mays* mesophyll chloroplast, x18,500.

Plate 21e and f Further details of *Zea mays* grana are illustrated here in top view (21e, x55,000) and side view (21f, x75,000). The grana contain many more discs than those in 21c, and the stroma (S) is much subdivided by frets (F) passing between the grana. A few ribosomes (circled) are seen in the stroma. In 21e the frets are in oblique view, appearing merely as grey shadows (at white arrowheads). The number of striations running across the grana in (e) increases with increasing obliqueness of section angle to the grana stack. Each striation represents an oblique slice through one disc. The granum at top right has been sectioned exactly in the plane of the discs and shows numerous frets. Since the section is only thick enough to accommodate about 3 granum discs, this means that each disc must develop many connections to frets, as in the idealised diagram of a granum (below):

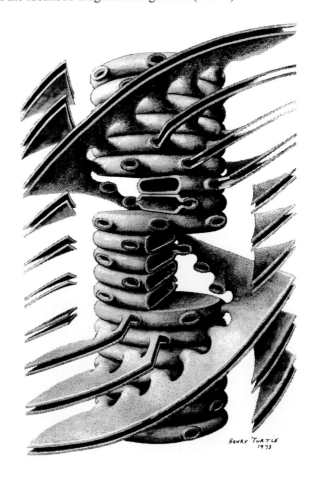

HENRY TURTLE
1973

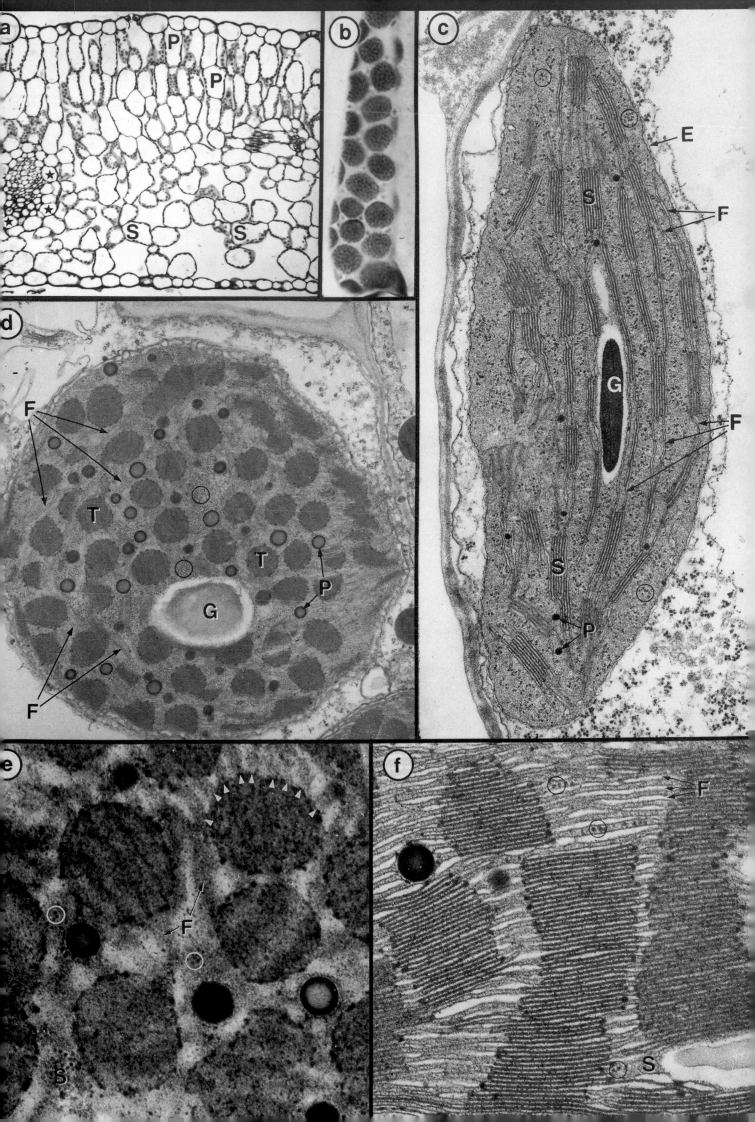

22 Plastids (3): Chloroplast Membranes

Introduction: The light reactions of photosynthesis commence with absoption of light energy by pigment-protein complexes embedded in thylakoid membranes. These light-harvesting pigments absorb over a wide range of wavelengths. They do not themselves process the absorbed energy: their role is to act as antennae and to pass the energy of excitation to *reaction centres* which contain special forms of chlorophyll capable of transferring the received energy to electrons, which then enter an electron transport chain. In *photosystem II* the electrons come from water molecules (at the same time liberating oxygen and leaving the intra-thylakoid space relatively enriched with protons); in *photosystem I* the electrons come from a long-wavelength absorbing form of chlorophyll-*a* in the reaction centre. Electrons from photosystem II replenish those removed by photosystem I and add to their energy. The combined energy boosts are sufficient to drive reduction of NADP, liberating $NADPH_2$ into the stroma along with ATP. Both cofactors are used during photosynthetic reduction of CO_2.

Most of the protein of thylakoid membranes is in the form of light-harvesting complexes, photosystems I and II, and ATP-synthases. A great deal is known about the molecular organisation of these membrane complexes and the electron carrier molecules that move between them. Some lie at the inner (intra-thylakoid) face of the membrane, others are within the membrane and others are situated at the stromal face. On a larger scale, photosystem I and ATP synthases are concentrated where the thylakoids are in contact with the stroma. By contrast, photosystem II is concentrated in grana, where adjacent discs lie appressed to each other, out of contact with the stroma and largely separated from photosystem I.

Grana are virtually ubiquitous in higher plants and many green algae, but it is not clear what special functions their structure confers. They are especially prominent in shade plants, as compared with plants that tolerate high light intensities. They can be "unstacked" and "restacked" *in vitro*. *In vivo* they are probably held as stacks by intermolecular forces between chlorophyll a- and b-protein complexes in photosystem II, operating across the narrow gap between adjacent membranes, and by cationic bridges. Shade plants have more of the light-harvesting chlorophyll a- and b-protein complexes, correlating with their more extensive grana stacks. Remarkably, when the membranes are "unstacked" *in vitro*, the various macromolecular complexes become randomised by diffusion, but can then resegregate again if the membrane is "restacked". One suggested function for granal architecture concerns damage to photosystem II by excess light ("photoinhibition"). One of its protein constituents (D1-protein) is especially prone to breakdown. Repair is by resynthesis of D1 on chloroplast ribosomes and its insertion into stroma thylakoids. Appressed disc regions may be biochemical havens where

damaged photosystem II units can continue to absorb excess light and thus serve to protect still-functional units while awaiting replacement D1. Another suggestion concerns reaction kinetics: photosystem I traps and processes electrons faster than photosystem II, so the argument is that it has to be segregated because it would harm photosystem II by bleeding electrons away from it if the two photosystems were allowed to intermingle.

The freeze-fracture electron micrographs shown here illustrate the complex sidedness and regional differentiation of thylakoids, in the form of particle sizes and distributions. Unfortunately, it is not yet possible to put functional labels on all of the particles that are seen.

Plate 22a This ultra-thin section of part of a lupin (*Lupinus* sp.) leaf chloroplast shows the chloroplast envelope (E) and grana and stroma thylakoids. A narrow gap between adjacent granum discs is visible where the plane of the section is at right angles to the plane of the discs (white arrowheads). This is where the stacking forces operate. Continuity between fret channels and the lumen of the discs is visible (open arrows). x140,000.

Plate 22b Grana of shade plants have very numerous discs, examined here by freeze-etching. The fracture plane has passed down two grana (G, lower left and upper right) approximately at right angles to the plane of the discs. A third granum, in the centre of the micrograph, lay obliquely to the fracture, so that surface views of membranes are exposed. The clear white bands are steps where the fracture descends from one membrane to another. The intervening particle-studded areas represent the interior of the successive disc membranes. One face carries large particles (called the "B-face"; it backs on to the *lumen* of the thylakoid), the other face has larger numbers of smaller particles (the "C-face", which backs on to the *stroma*, or, in grana, the *adhesion face* between adjacent discs). On passing from the discs to the frets that interconnect the grana, the large B-face particles become less frequent (arrows from labelled B-face). They may therefore represent some portion of photosystem II complexes. The small C-face particles do not diminish in numbers from disc to fret (arrows from labelled C-face). *Alocasia* (a shade plant), x67,000.

Plate 22c In this high magnification view, about two thirds of the area of a granum disc membrane is seen in the bottom part of the picture, showing scattered large, putatively photosystem II components typical of the B-face. Frets radiate away from the circumference of the disc, and the fracture has exposed examples of both their B- and their C-faces (B and C respectively). Several fret B-faces are continuous with the B-face of the disc (arrows), and as in (b), the frequency of the larger particles diminishes, moving away from the disc. By contrast, the fret C-faces have very numerous small particles. *Lomandra longifolia* (a shade plant), x105,000.
(b) and (c) were kindly provided by D. Goodchild.

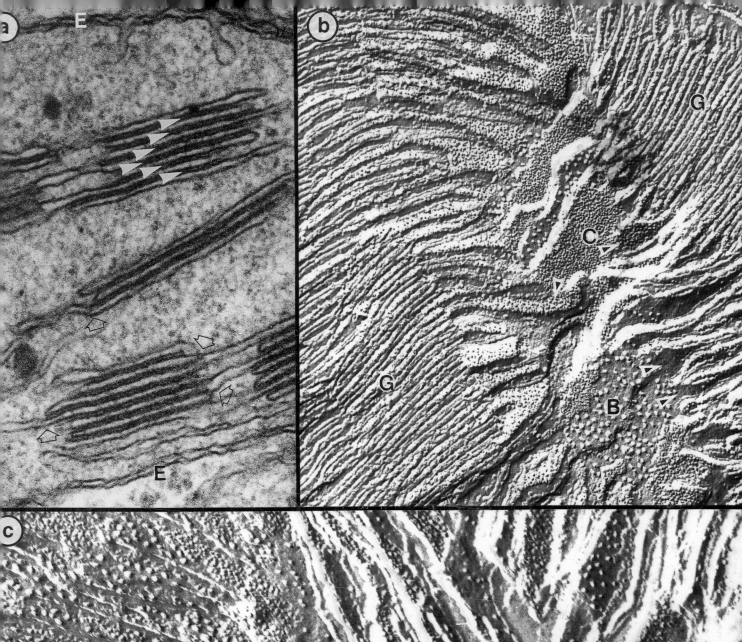
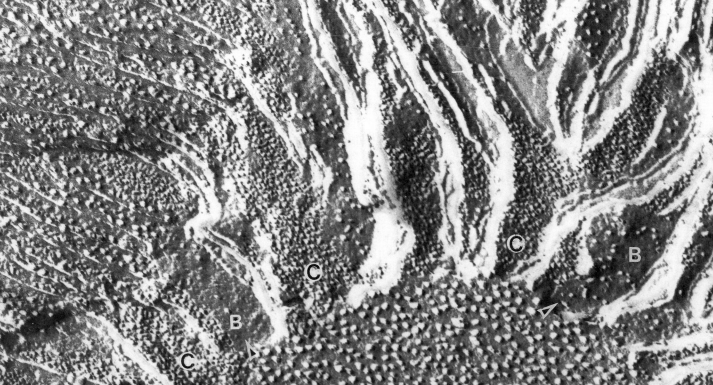

23   Plastids (4): Components of the Stroma

Introduction: The light reactions of photosynthesis release ATP (chemical energy) and $NADPH_2$ (reducing power) into the chloroplast stroma, where they drive the "dark" reactions of the *photosynthetic carbon reduction cycle (Calvin cycle)*. A CO_2-acceptor molecule with 5 carbon atoms, ribulose 1,5-bisphosphate, undergoes carboxylation to yield two 3-carbon intermediates. The ATP and $NADPH_2$ then engage in sequential reactions that reduce the incorporated CO_2 to the level of carbohydrate ($[CH_2O]_n$). Further reactions and continued turns of the cycle regenerate the 5-carbon acceptor (which is why the sequence is called a cycle) and produce three carbon sugar phosphate, the net gain of the system. This product of the cycle can be used to make starch in the chloroplast stroma, usually as a temporary store, later to be exported. Alternatively, it can be transported across the chloroplast envelope to the cytoplasm, and there converted to sucrose. Sucrose is the major form in which carbon is moved around plants, *via* plasmodesmata and in the phloem, to sites of growth or storage.

The most abundant of the Calvin cycle enzymes is the one that adds CO_2 to the acceptor. It is called ribulose bisphosphate carboxylase-oxygenase, or *rubisco* for short. A large protein (mol wt 0.5×10^6 Da), it is highly significant because it is the main enzymatic gateway for entry of carbon into the organic world. Compared with most enzymes it operates very slowly, i.e. it catalyses relatively few carboxylation reactions per unit time. To keep pace with the photoreactions of photosynthesis it has to be present in bulk - in fact it is not merely the most abundant protein in the chloroplast stroma, but the most abundant protein on earth. However its catalytic site also has a second enzymatic property, the "oxygenase" component, which competes with photosynthesis by dissipating carbon. Plants have evolved several systems that alleviate this serious detriment to plant productivity, two of them illustrated in Plates 24 and 30.

Plate 23a   Components of the chloroplast stroma are shown in this micrograph of a greening oat leaf:-

Nucleoid: The nucleoid (N) sectioned here contains fine fibrils, 2-3nm in thickness (small arrows), the histone-free DNA of the plastid. This DNA is present in several copies and carries about 120 genes, including multiple copies of genes for plastid ribosomal RNA. The associated granules (double arrows) may in fact be developing plastid ribosomes (also Plate 19b).

Ribosomes: Plastid ribosomes may be single or in polyribosome chains (rectangle) or clusters (between stars). Some lie in contact with the thylakoids (arrowheads) probably linked to them by protein chains that are being inserted into the lumen or the membrane, like the attachment of polyribosomes to endoplasmic reticulum. As seen in other plates, chloroplast ribosomes are smaller than cytoplasmic ribosomes, present at lower left outside the plastid envelope (E).

Proteinaceous ground substance: The general background in the plastid stroma is finely particulate (except in the nucleoid, which excludes the rest of the stroma). From the quantities known to be present, many of the particles must consist of rubisco, some molecules of which are shown negatively-stained (pale against a dark background) at high magnification in the inset at top left (one example ringed). Each molecule is a particle of side about 10nm. As diagrammed in the second inset, higher plant rubisco consists of 8 identical large subunits in four pairs (these are coded by plastid genes and are made in the plastid; they carry the catalytic sites, diagrammed as white spots) and 8 identical small subunits in two rings of four, sited above and below the ring of large subunits (the small subunits are coded by nuclear genes).

The bundle of fibrils (S) occurs naturally in plastids of most species of *Avena*. There is good evidence that it consists of aggregates of a specific β-glucosidase enzyme and it is one example of the numerous chemical defence mechanisms that have been evolved by plants. If the cell membranes are broken down by entry of a fungal hypha, the glucosidase enzyme gains access to its substrate (a steroidal saponin called avenacoside) which is normally stored separately, in the vacuole. A fungitoxic product is then liberated by the action of the glucosidase on its substrate. x94,000, inset x430,000.

Plate 23b   Iron is sometimes stored in the plastid stroma, in the form of the protein *ferritin*, occasionally in large masses, as in this lupin leaf chloroplast. Each electron dense point represents a core of several hundred iron atoms in the centre of a protein molecule. The protein shell around the iron core is not visible here. When ferritin crystallizes its molecules are arranged as in the inset. x120,000.

Plate 23c-g   Chloroplasts of most algae contain one or more *pyrenoids*, which are concentrations of rubisco (though rubisco is also present elsewhere in the stroma). Pyrenoids are conspicuous in green algae because they develop a sheath of starch grains. This sequence of pictures illustrates changes that occur during one day-night cycle in the life of a *Chlorella* cell. In very young cells (c) starch (solid stars) is synthesised as thin plates around the pyrenoid (P) and accumulates while the cell photosynthesizes and grows during the day (d,e). Starch grains also appear in other parts of the chloroplast stroma (open stars in d, e and f) but the pyrenoid starch is especially labile. It is consumed in the dark period before the non-pyrenoid starch as the cell begins to divide (f). The products of cell division generally have starch-free pyrenoids as they begin to photosynthesise at the start of the next cycle (g).

(c) and (d) illustrate (as does Plate 18d,e) the proximity of the single microbody (M) to the pyrenoid in this alga. All x33,000, except (f), at x18,000. Kindly provided by A. Atkinson, Jr.

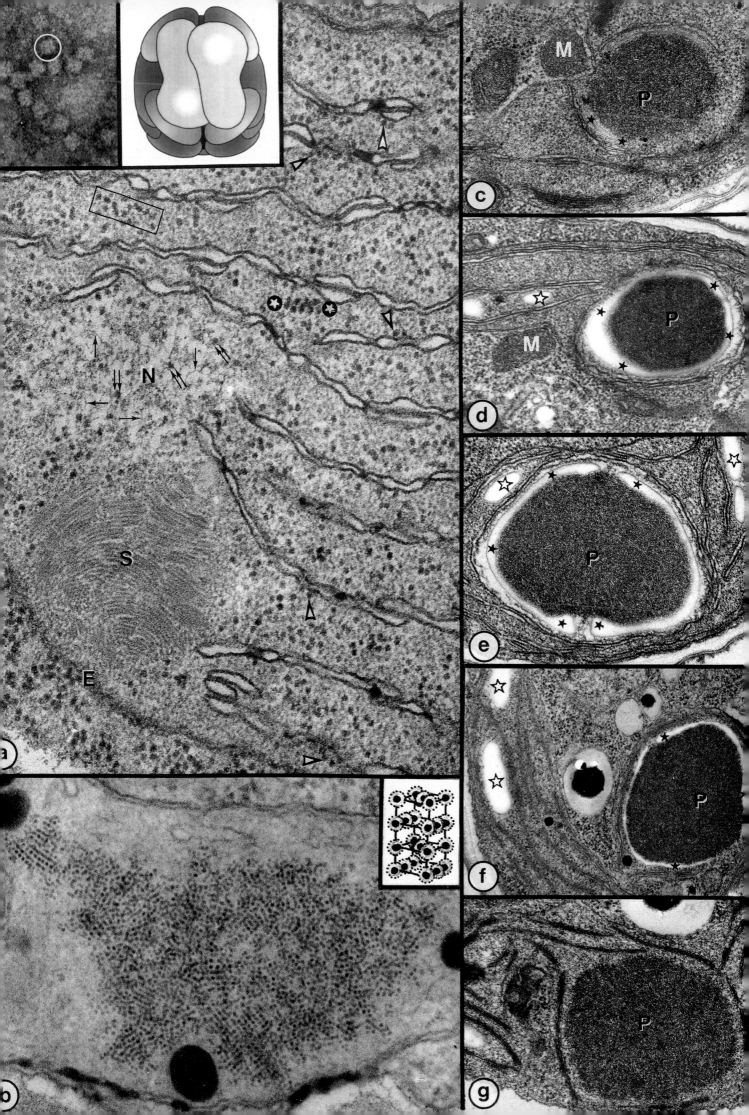

24 Plastids (5): Dimorphic Chloroplasts in the C-4 Plant, *Zea mays*

Introduction: Ribulose bisphosphate carboxylase-oxygenase (rubisco), the CO_2-fixing enzyme of photosynthesis, probably evolved when the earth's atmosphere was lacking in oxygen but rich in CO_2. Conditions thus favoured rubisco's carboxylation reaction. Its second catalytic activity, which is present now and presumably was also present then, was of little or no consequence because it required oxygen. However, as photosynthesis released more and more oxygen into the atmosphere, rubisco's second, oxidative, catalytic reaction became more serious. It breaks down the vital 5-carbon CO_2-acceptor, ribulose bisphosphate, and some of the carbon is then lost by decarboxylation. For unknown reasons, evolution of more efficient, solely carboxylative, forms of rubisco has not occurred, although the subunit composition (see Plate 23a) and the ratio of the two enzyme activities have changed, judging by differences between the enzyme as found now in plants and bacteria.

However, plants have evolved mechanisms to reduce the loss of carbon through substrate oxidation by rubisco. One mechanism, found in "C-3" plants (which rely solely on the Calvin cycle type of dark reactions), limits losses by reactions involving chloroplasts, mitochondria and microbodies (see Plate 30). Two other mechanisms evolved independently in several groups of flowering plants. They enhance the basic Calvin cycle by increasing the concentration of CO_2 around rubisco, thus favouring productive carboxylation over dissipative losses by sequential oxidation and decarboxylation:-

The first mechanism separates CO_2 absorbtion from CO_2 reduction temporally. In "CAM" plants (mostly succulents; the acronym stands for Crassulacean acid metabolism) CO_2 enters leaves at night through open stomata and is stored in mesophyll cell vacuoles as the 4-carbon compound, malic acid. Then in the daytime the stomata close and the stored CO_2 is released under low oxygen conditions for efficient re-fixation by rubisco in the *same* photosynthetic tissue.

In the second mechanism, found in "C-4" plants, a similar result is achieved using dimensions of space rather than time. Their mesophyll cells contain chloroplasts that lack rubisco but still use light energy to make ATP and $NADPH_2$. CO_2 is fixed and reduced in the cytoplasm into 4-carbon organic acids (hence the "C-4" acronym) by enzymes that are not sensitive to oxygen. The 4-carbon acids (malic or aspartic, depending on the type of C-4 plant) diffuse from cell to cell through plasmodesmata into bundle sheath cells surrounding the leaf veins, whose chloroplasts *do* contain rubisco. There the acids are broken down to release CO_2 for re-fixation by rubisco and subsequent processing by the Calvin cycle (as in CAM plants). In effect, C-4 plants enrich their bundle sheath with CO_2, making local conditions more favourable for rubisco's carboxylation than pertain in the mesophyll. The "C-4 syndrome" also includes the important feature of high efficiency in using water (less water transpired per unit of carbon fixed than in C-3 plants). C-4 species are thus suited to hot, dry climates, and some, like maize, are very important in agriculture.

Plate 24a The dimorphic chloroplasts found in C-4 plants are distinguishable even in the light microscope, as here in a section of a maize (corn) leaf. The xylem (X) and phloem (P) of each vein are surrounded by bundle sheath cells, and these in turn by mesophyll cells. Grana are darkly stained in the mesophyll chloroplasts (e.g. large arrowheads). Starch grains are unstained (small arrows), and are abundant by-products of the Calvin cycle in the *agranal* bundle sheath chloroplasts. x1750.

Plate 24b Bundle sheath (BS, lower right) and mesophyll (M, upper left) chloroplasts of maize are compared in this electron micrograph. The former contain starch grains (G) in the stroma between the simple, agranal, internal membranes. The latter possess grana and frets (seen in more detail in Plate 22). The leaf was still growing when it was fixed and the chloroplasts had not completed their development, as indicated by the presence of a small region of prolamellar body lattice (solid arrow), perhaps a relic of membrane growth in darkness during the night before the early morning harvesting of the material (see also Plates 25,26). The cell wall passing diagonally across the micrograph contains a layer of suberised material that surrounds each bundle sheath cell (open arrows). Plasmodesmata (P, parts of two groups included) interconnect bundle sheath and mesophyll protoplasts and are important avenues of transport of intermediates between the two tissues. The tonoplast (T) of each cell can be seen, separated from the chloroplasts by a thin layer of cytoplasm. x27,000.

Plate 24c Parts of bundle sheath (BS) and mesophyll (M) chloroplasts are shown here at higher magnification. The intervening wall contains a suberized lamella (obliquely sectioned, open arrows) and plasmodesmata (P). Mesophyll chloroplasts have no rubisco but possess normal grana. Photosystems I and II operate to produce ATP and $NADPH_2$, but they are used to fix CO_2 into 4-C acids by a cytoplasmic enzyme, rather than in Calvin cycle reactions. By contrast, thylakoids in bundle sheath chloroplasts are largely unstacked, with diminished photosystem II activity. They can still make ATP, which they use along with reducing power obtained from the incoming 4-C acids to drive the Calvin cycle. Note the economy of this dimorphic chloroplast development - the plant does not invest resources in making rubisco (mesophyll) or photosystem II (bundle sheath) where these protein-rich (and therefore expensive) systems are not required. Both types of chloroplast invaginate the inner envelope membrane (large solid arrowheads) to form *chloroplast peripheral reticulum,* with perforations penetrated by the stroma (small arrows). This arrangement is not exclusive to C-4 plants. x64,000.

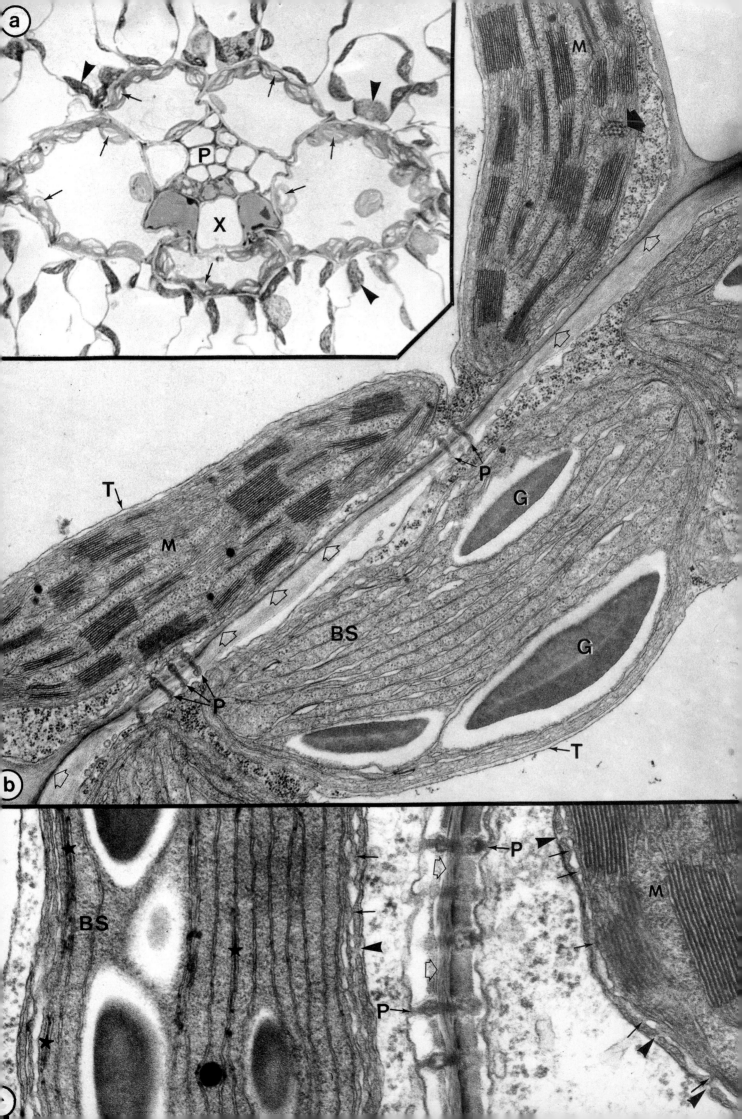

25 Plastids (6): Etioplasts and Prolamellar Bodies (i)

Introduction: When plants are grown in darkness they develop a set of symptoms known as *etiolation*. Their shoots become very elongated, resembling those of seedlings germinating underground. New leaves are white or yellow coloured and rarely produce green chlorophyll pigments. Upon return to the light, normal growth resumes and the *etiolated* leaves that were formed in the dark turn green and begin to photosynthesise.

Plastids undergo limited development in etiolated leaves, changing from the proplastid precursor condition into a special state, the *etioplast*. Etioplasts are not intermediates in normal chloroplast development, as seen, for example, in leaf primordia growing in the light (Plate 20). Rather they are a diversion to a unique arrested state, brought about by lack of light. The proteins of mature chloroplasts are absent or else present in very low amounts. Etioplasts have some of the lipids of chloroplasts but no chlorophyll. Instead, they contain a precursor pigment called *protochlorophyllide*, and a remarkable membrane system known as the *prolamellar body*, illustrated in this and the next plate.

The prolamellar body apparently forms as a result of continued synthesis of thylakoid lipids in darkness without parallel synthesis of thylakoid proteins. It is not known how the preponderance of lipids (about 75%) determines the peculiar morphology of the membranes. They consist of tubes that branch in three dimensions and form a quasi-crystalline lattice whose continuous surface is curved in opposite senses at every point. The surface is also in contact with the stroma at every point, and likewise the compartment inside the tubes is continuous. Prolamellar body membranes are examples of *periodic minimal surfaces*, geometric forms consisting of potentially infinite arrays of repeating units. They have long been known to mathematicians but here are generated naturally by forces of surface tension acting on a special type of biological membrane.

Protochlorophyllide is a key compound in etioplasts, although it is present at only a few percent of the quantity of chlorophyll that is found in a mature chloroplast. It is in part responsible for the arrested state of development exhibited by etioplasts, because it inhibits an early step in biosynthesis of chlorophyll-type pigments. Thus accumulation of some protochlorophyllide inhibits its own further production. Synthesis resumes if it is removed by conversion to chlorophyll, for which two enzymatic steps are needed. These steps do not occur in etioplasts because the first one needs light. When illuminated, protochlorophyllide triggers its own removal by absorbing photons to reach an excited state. This allows it to be reduced by an $NADPH_2$-requiring enzyme that is made in the cytoplasm but becomes the predominant protein of the prolamellar body. In contrast to what happens in dark-grown plants, protochlorophyllide is continually reduced during growth in continuous light, and so never reaches high enough concentrations to inhibit the overall process of chlorophyll synthesis. Chloroplast development in the light therefore bypasses the etioplast stage. The second enzymatic step fundamentally changes the way the pigment associates with lipids and proteins by adding a 20-carbon hydrocarbon chain to the molecule (until this addition, the names of the immediate precursor pigments have the suffix *ide*). However, light does more than trigger reduction of protochlorophyllide. It also activates genes for production of chloroplast proteins that have not been made in quantity in the dark. These light-induced "greening" reactions are described in Plate 27.

Plates 25, 26 are from dark-grown oat seedlings (*Avena sativa*). Legends for Plate 25 e-h are on Plate 26.

Plate 25a In this low magnification survey micrograph, two etioplasts display the two membranes of the plastid envelope (black arrow), and numerous plastid ribosomes. The prolamellar bodies (PLB) can be seen to be semi-crystalline lattices of membrane, but in these examples the plane of section does not display regular lattice planes to advantage. Subsequent micrographs have been selected for this purpose. The lattice at the bottom centre region of the left hand prolamellar body (above the white arrow) is obviously different from the neighbouring lattice and is shown in more detail in Plate 26h. x47,000.

Plate 25b, c, d The simplest form of prolamellar body lattice is illustrated first, although it is uncommon, by a micrograph (b), a 3-dimensional drawing (c), and a model (d). It is composed of tubes, branched and interconnected in three axes at right angles to one another to form a cubic lattice. The unit of construction is shown enclosed within the square on the micrograph, and in the left hand diagram (below). The etioplast stroma penetrates the prolamellar body between the tubes, but not much structure is seen in the stroma component because uranyl acetate stain (which adds contrast to ribosomes) was not used (cf. later images). Models of the type shown in (d) will be used to illustrate the other types of prolamellar body (below): comparison with (c) emphasizes that such models represent only the orientation of the tubes, and not their diameter and smoothly confluent surface contours. (b) x78,000.

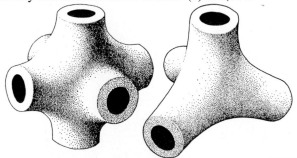

unit of cubic lattice   unit of tetrahedral lattice

(continued on next plate)

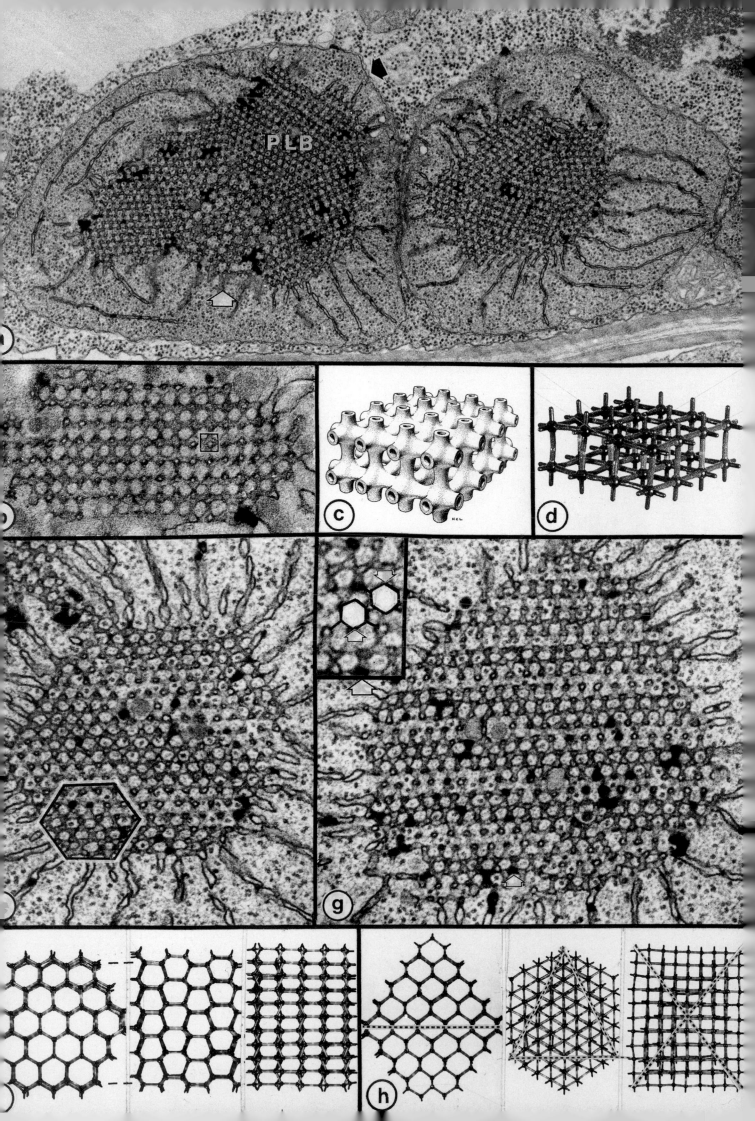

26 Plastids (7): Etioplasts and Prolamellar Bodies (ii)

(continued from previous plate)

Plate 25e, f Most prolamellar body membrane lattices consist of tetrahedrally-branched, tubular repeating units (see diagram on previous plate), smoothly interconnected in three dimensions. Two crystallographic symmetries develop, analogous to the lattices of two related zinc sulphide minerals. One form, shown in (e) and (f), is analogous to the mineral *wurtzite*. The view of the model in 25f (left hand side) matches an area such as that marked on the micrograph. The other views of the model are side views - looking at an edge (centre) or at a face (right hand side) of the hexagonal "crystal". This and other micrographs on the Plates 25 and 26 show many etioplast ribosomes in the stroma component of the lattice. They also show continuity of prolamellar body lattice tubes and outwardly-projecting, flattened *prothylakoids*. x70,000.

Plate 25 g, h This micrograph and models show the alternative tetrahedral lattice form, analogous to bipyramidal crystals of the "zincblende" form of zinc sulphide (or to the lattice of carbon atoms in diamonds). The views of the model show an edge (left), a face (centre) and a vertex (right). This lattice differs from 25(e,f) in that successive sheets of hexagonal rings are out of register. The centre picture of the model illustrates the overlapping hexagons of the lattice. The section (g) is somewhat tilted relative to (h) (centre), but individual hexagons are seen in horizontal bands, each band being part of one plane of the lattice. The inset (of the area above the arrow at lower centre of the main picture) shows successive bands, out of register by half a hexagon (unlike the "in-register" wurtzite lattice in 25e,f). x70,000, inset x120,000.

Plate 26 a Tetrahedrally branched models of carbon atoms were used (as in Plate 25f, h) to construct this large tetrahedron-shaped complex, made up of very numerous individual tetrahedral units. It is relevant because it is found as a large-scale building block in many prolamellar bodies, e.g. 26c,d,g. It consists of 'zincblende' type lattice (as Plate 25g,h) but with a pentagonal dodecahedron at each vertex, and special 5- and 6-membered rings running along each edge. The latter arise when these large building blocks are joined to one another in complex prolamellar bodies. A pentagonal dodecahedron is diagrammed below to show how it is made up of standard tetrahedral membrane repeating units.

Plate 26 b, c, d, e *Centric* prolamellar bodies have 20 large tetrahedral complexes radiating outwards, one from each of the 20 vertices of a central pentagonal dodecahedron, visible in section at the centre of (c) and (d). An icosahedral-shaped prolamellar body is generated, as seen in (b), where a model is viewed looking straight at one vertex (the edges of the tetrahedral complexes on the near side of the model have been drawn in white). It

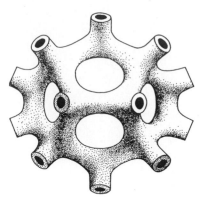

tetrahedral units joined in a pentagonal dodecahedron

is impossible to see all 20 large complexes in any one section. In the approximately median sections (c) and (d), parts of 10 large tetrahedral complexes are visible, each one radiating out from a vertex of a central pentagonal dodecahedron. Facets of the 10 sectors are marked by straight white lines. For comparison, a median slice of the model is shown in (e). (c) x38,000; (d) x73,000.

Plate 26 f, g Here the large tetrahedral complexes (26a) are combined in a more complex fashion - not just radiating out from a single central pentagonal dodecahedron, but as building blocks filling space between *many* evenly spaced pentagonal dodecahedra. 26f represents a slice through a model equivalent to the ultrathin section seen in (g). One pentagonal dodecahedron is at the centre, 4 others are marked (arrows) at the periphery, and yet others were above and below the plane of this slice. Outlined triangular sectors in (g) represent triangular faces of complexes of the type shown in (a). Their counterparts in a similarly-oriented model are seen in (f). The prolamellar body in (g), although more extensive than the model, is still incomplete. If it were to enlarge further it would place pentagonal dodecahedra at positions marked by asterisks. (g) x84,000.

Plate 26 h, i, j, k Symmetry in this very complex form of prolamellar body is hard to see because the sections are thinner than the lattice spacings. An area comparable to (j) is marked on the micrograph (h). The lattice is mostly composed of rows of pentagonal dodecahedra, interconnected at 60° to one another with intervening gaps which create the hexagonal pattern seen in (j). Rows inclined at 60° to one another can be seen if the page is viewed at a shallow angle along the three axes parallel to the sides of the white hexagon. Side views of the lattice (edge-on in (i) and face-on in (k)) show how the successive strata of pentagonal dodecahedra (between brackets) are connected. This type of prolamellar body can occur in isolation, or it may be joined to one of the other types (as in Plate 25a). In some plants it is common early in etiolation but gives way to the more compact types if time in darkness is prolonged. (h) x70,000.

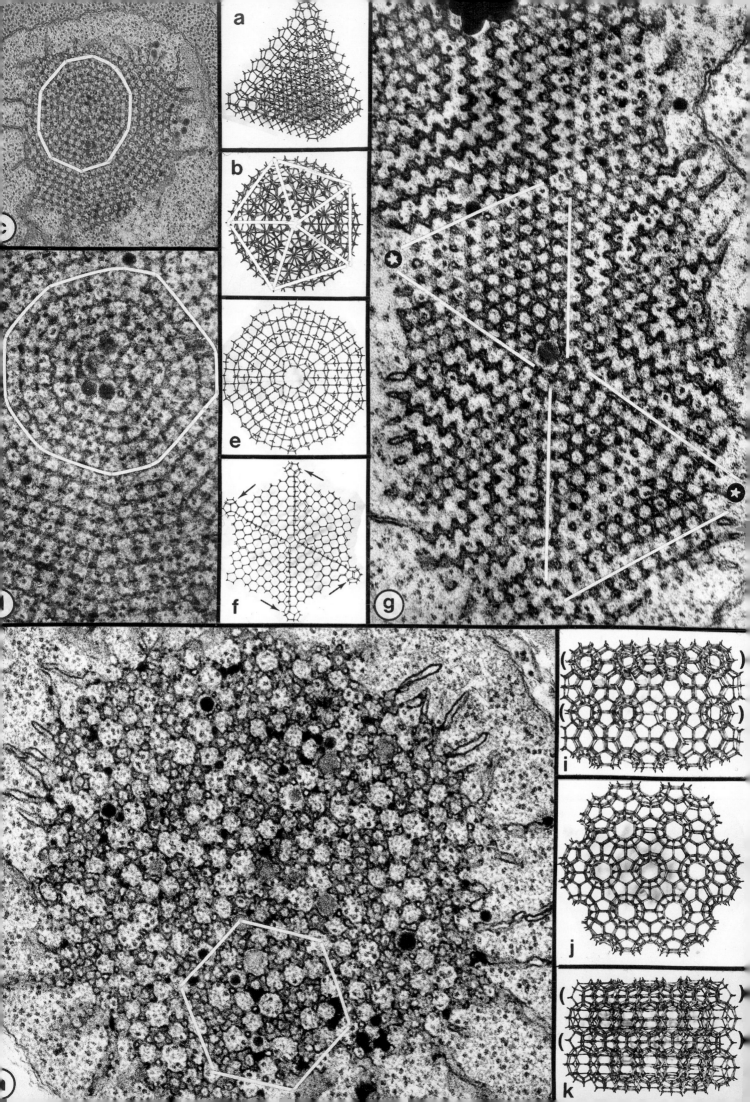

27 Plastids (8): The Greening Process: From Etioplast to Chloroplast

Introduction: Plate 20 shows two of the possible pathways of development from proplastids: (i) to chloroplasts, in plants grown in the light, and (ii) to etioplasts, in plants that are kept in darkness. The present plate shows the pathway of conversion of etioplasts to chloroplasts, when a dark-grown plant is illuminated (the *greening process*).

Illumination very rapidly triggers conversion of the protochlorophyllide in the prolamellar body to chlorophyll (see Plate 25). The prolamellar body concomitantly loses its crystallinity and its membrane undergoes metamorphosis into flattened primary thylakoids. Slower changes are also set in train. Over a period of one or two days additional membrane is assembled into grana and stroma lamellae. Photosynthetic pigment complexes, enzymes, electron transport molecules and plastid ribosomes are made and inserted into either the developing thylakoids or the stroma. These activities involve numerous genes in the plastids and nuclei, and also, for those substances made in the cytosol, transport across the chloroplast envelope membranes.

Comparable developments occur in the direct pathway from proplastids to chloroplasts (in the light). There, photosystem I, photosystem II and Calvin cycle activity all appear concomitantly in leaf samples, perhaps because each sample contains vast numbers of plastids, spanning many different stages of development. By contrast, in the pathway from etioplasts to chloroplasts, abrupt illumination triggers unusually synchronous development of plastids throughout the leaf, and the subsequent acquisition of photosynthetic functions is stepwise. Notably, photosystem I becomes operational early in greening, considerably before photosystem II, which does not become active until appressed discs in grana begin to appear, several hours after greening starts (see Plate 22).

Protochlorophyllide is clearly the photoreceptor for its own conversion to chlorophyll (see Plate 25). However, there is another important photoreceptor, *phytochrome,* in plants. It is responsible for initiating most light-induced responses, including the de-etiolation syndrome and other aspects of chloroplast development. It exists in two forms. That found in dark conditions is converted by red wavelengths to the second. The latter is capable of activating many genes, both in nuclei and in plastids. Such light-activated genes are known as *photogenes.* Prominent examples in greening reactions include the genes for the large and small subunits of rubisco and the gene for the thylakoid protein that binds chlorophylls to make light-harvesting complexes.

As in Plates 20, 25 and 26, the micrographs in this plate show plastids in oat (*Avena sativa*) leaves. Each left-right pair shows developing thylakoids in profile view (a,c,e) and in face view (b,d,f) at three different times in the greening process.

Plate 27a, b Two hours after the dark-grown plants were brought into the light, a prolamellar body remnant (PR) has not completely dispersed, though the regularity of its lattice has been lost (in fact it was lost within seconds of the onset of illumination). Many of the perforations that were delimited by tubular membranes in the former prolamellar body still survive as small pores through the primary thylakoids (small arrows in the profile (a) and face (b) views of the thylakoids). The plastid ribosomes, formerly largely single, are now mostly in clusters and chains, suggestive of polyribosomes (e.g. open arrows) - a sign that gene transcription and protein synthesis has been activated in the greening plastid. Invaginations of the inner plastid envelope membrane (stars) and a nucleoid area (N) are seen in (a).

Plate 27c, d The amount of protochlorophyll that is converted by light to chlorophyll is very small compared with the amount of chlorophyll found in a mature chloroplast. Net synthesis of chlorophyll begins in the greening plastids after a lag period, which in the material shown here lasts 2-3 hours, and during which no new chlorophyll, and probably no new membrane, is produced. Some photosystem I activity is present at this time. The beginning of the period of rapid synthesis of membrane is marked by the appearance of photosystem II and of portions of membrane overlapping the primary thylakoids (large arrowheads). This is the first stage of granum formation, shown in both profile (c) and face (d) view (and at higher magnification in Plate 23a). One of the overlapping discs is sectioned through its fret connection (opposed arrowheads). More of the perforations seen in (a) and (b) have by now disappeared from the primary thylakoids, though a large prolamellar body remnant still survives (right hand side of (c)). The membranes of the primary thylakoids are still continuous with those of the prolamellar body (e.g. at small arrows).

Plate 27e, f After 10 hours in the light the leaves are obviously green, but still not as green as a leaf that contains mature chloroplasts. Chlorophyll and membrane synthesis has progressed but more light-harvesting chlorophylls will be added. The small membrane overlaps seen in (c) and (d) have extended to become full-sized, multiple granum discs (see (f)), interconnected by frets derived from the primary thylakoids (see (e)). A mass of plastoglobuli marks the remnant of the prolamellar body in each picture. Figures (c), (d), (e) and (f) all show that chains of ribosomes lie free in the stroma and on the surface of the growing thylakoid membrane (beside asterisks in all four micrographs). The latter resemble polyribosomes on rough endoplasmic reticulum membranes. These thylakoid-associated polyribosomes may be synthesizing protein molecules which pass directly into the growing membrane surface or into the thylakoid lumen.

Magnification x36,000 in all except (c), at x64,000.

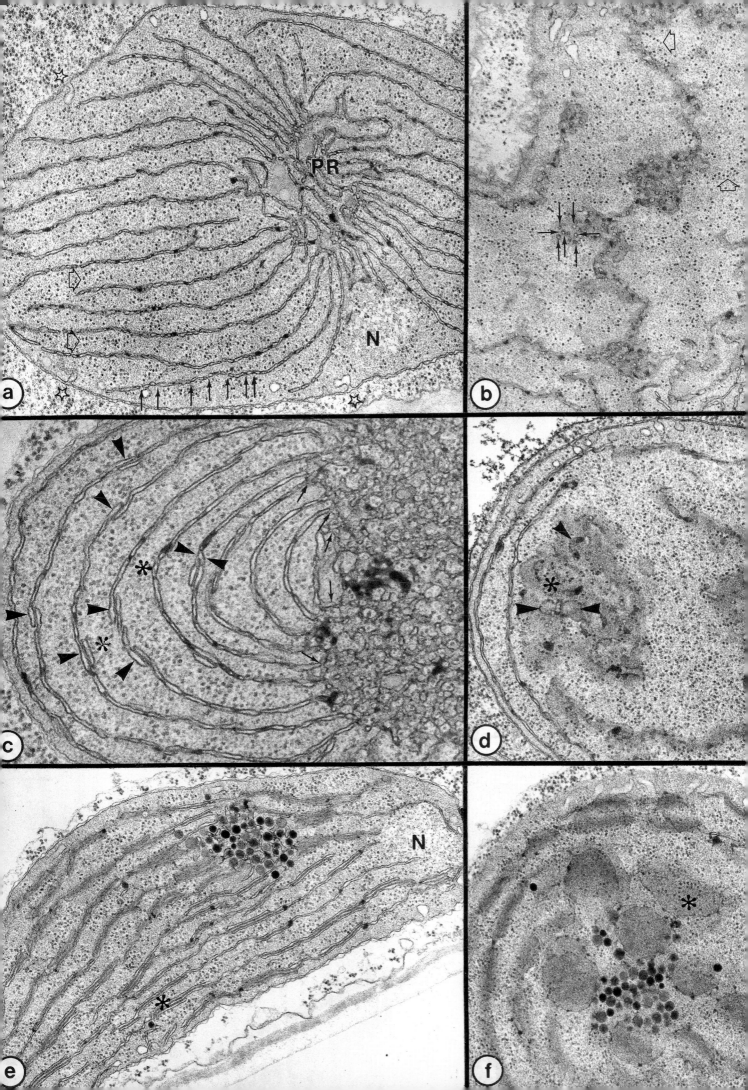

28   Plastids (9): Amyloplasts

Introduction: The immediate product of the Calvin cycle, 3-carbon sugar phosphate, can be used directly to make starch in the chloroplast stroma (see Plate 23c-g for an example), or it can be exported to the cytoplasm to be made into sucrose. The sucrose can be transported around the plant in the phloem, and can be consumed in growth (etc.) or stored. This plate shows a form of plastid, the *amyloplast*, which specialises in storage of starch manufactured from translocated sucrose, and is found in tissues that store food reserves (see also Plate 14a). They lay down such large starch grains that the plastid stroma is reduced to a thin layer between the grain and the enclosing envelope membranes. Starch is stored inside plastids in all plants except red algae, which make it in the cytoplasm. One advantage of storing carbohydrate as starch is that, unlike the soluble, low molecular weight sucrose, it is not osmotically active.

The prefix "*amylo-*" relates to the names given to the two main types of starch, *amylose* (long unbranched chains of glucose units) and *amylopectin* (occasionally branched chains). The ratio of the two forms and the extent of branching varies from one species of plant to another, and even within different organs of the same plant. Certain starches, called *waxy*, are almost entirely amylopectin. The extreme of branching is reached in the plant *Cecropia*, which develops food bodies that are harvested by ants; in these (but not in the rest of the plant) the plastids synthesise *glycogen*, the equivalent polysaccharide reserve in fungal and animal tissues. Variations in starch structure are commercially important in the food and brewing industries.

A transport protein in the chloroplast envelope whose function it is to transport 3-carbon sugar phosphate from the stroma to the cytosol has a key role, not just in fueling sucrose production in the cytoplasm, but also in regulating photosynthesis itself. The reason is that the phosphate component of the exported sugar phosphate must be returned to the chloroplast stroma to enable the Calvin cycle to produce *more* sugar phosphate. The transport protein in fact exchanges sugar phosphate (moving outwards) for phosphate ions (moving inwards) in a strict one-to-one ratio. The phosphate that replenishes the Calvin cycle is released from the exported sugar phosphate when it is converted to sucrose in the cytoplasm. Starch formation in the stroma also releases phosphate. Thus, conversion of the immediate product of photosynthesis to sucrose and/or starch actually promotes continuation of carbon fixation, mediated by phosphate ions and their transport across the chloroplast envelope.

Plate 28a, b These two scanning electron micrographs show cells in a piece of potato tuber that was prepared by conventional fixation and then dried by the 'critical point' method, in which distortion is minimized. A freshly-cut surface was exposed before taking the pictures. The cells are highly vacuolate in life, indeed some parts of those in (a) appear to be nearly empty. The most conspicuous of the cytoplasmic components is the population of starch grains. The grains are ovoid when large, and nearly spherical when small. The cell at the bottom of (a) is seen at higher magnification in (b). Amyloplast envelope membranes are not resolved; they lie closely appressed to the starch grains. The largest grains are about 30 μm x 50 μm, and the smaller spheres about 3 μm in diameter. Numerous strands of cytoplasm, some with remains of more particulate cell components, form a network stretching out to a peripheral thin layer of cytoplasm at the cell wall. The strands presumably traversed vacuoles in the living state. (a) x280; (b) x680.

Plate 28c These three amyloplasts from a peripheral cell of a soybean (*Glycine*) root cap are very much smaller than the reserve amyloplasts in (a) and (b). Many starch grains (S) are present in each. General features of plastids that can be seen include the double membrane envelope (black arrow) and nucleoid areas (white arrow). The internal membrane system is not well developed, but stacked thylakoids occur occasionally (star). Thylakoids sometimes lie closely appressed to starch grains (open arrows), but it is not known whether this reflects membrane-activity in starch metabolism, or whether the starch grains merely became pressed against the membranes as they grew. x23,000.

Plate 28d The starch in sieve element plastids is of an unusual type, containing a high proportion of branched chains of glucose. Whereas ordinary amylase enzyme would digest potato starch grains, pretreatment with a special de-branching enzyme is necessary to break down sieve element starch. The grains are unusually electron-dense, and display a granular composition approaching that of glycogen (which is also a branched polymer of glucose). The double envelope can be seen around the upper grain. *Coleus* petiole sieve element, x15,000.

Plate 28e These amyloplasts grouped round the nucleus (N) in a young root cap cell of *Cosmea* each contain numerous starch grains (usually round in shape), and in addition accumulations of material that is extremely dense to electrons after processing for electron microscopy. The accumulations lie in distended intra-thylakoid compartments: in other words they are not in the stroma, where electron-dense plastoglobuli are found. In other material it has been found that this type of accumulation can be digested away from the section by treatment with the lipid-and protein-digesting enzymes lipase and pronase. It may therefore contain a lipoprotein. Phenolic material could also be present. Equally dense deposits are seen in the vacuoles (asterisks) and there is evidence (again from other material), that plastids can extrude phenol-containing droplets to the cytoplasm and vacuoles. Other features of the micrograph include mitochondria (M) with cristae and small dense granules, and lipid droplets (L). x26,000.

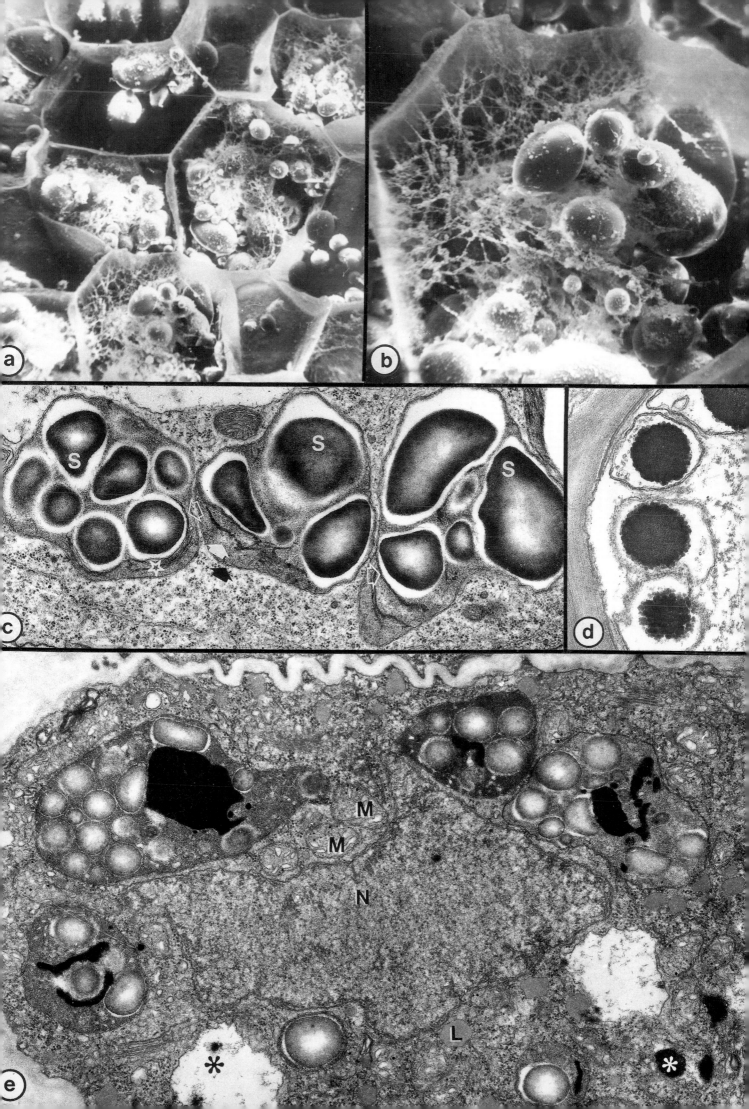

29   Plastids (10): Chromoplasts

Introduction: Although chlorophyll pigments dominate the colour of photosynthetic tissue (absorbing red and blue light but transmitting green wavelengths), other pigments are also present in chloroplasts. Carotenoid pigments are hydrocarbon chains with 40 carbon atoms. They absorb at the blue end of the spectrum and therefore appear yellow, orange or red, depending on their molecular structure. There are many of them, indeed they outnumber the chlorophylls in variety, though not in quantity. Xanthophyll pigments are oxygen-containing carotenoids, also present in chloroplasts. Two functions are recognised for these accessory pigments: they can absorb wavelengths that escape absorption by chlorophyll and pass the energy to the photosynthetic reaction centres; and they act as a photochemical safety valve, dissipating excess energy that might otherwise damage the photosynthetic apparatus when there is too much light (an important function described as protection against *photoinhibition*). Like chlorophylls, they occur complexed to proteins in thylakoids. This plate is concerned with a category of specialised plastid, *chromoplasts*, in which the chlorophylls are the lesser and the carotenoids the dominant pigments.

Chromoplasts are responsble for yellow, orange or red colours of many flowers and fruits and some roots. Familiar examples include daffodil flower petals and tomato fruit (both shown in this plate), citrus fruit, red peppers, carrot tap roots and sweet potato tubers. Colouration of plant tissues by chromoplasts is an alternative to colouration by storage of red/purple/blue water soluble anthocyanin pigments in vacuoles. Arrays of droplets of carotene pigments occur as *eyespots* within chloroplasts of many motile algae, but true chromoplasts evolved first in advanced forms of green algae (they occur in reproductive organs of members of the Charophyta) and are found throughout the groups of land plants.

Chromoplasts may arise directly from proplastids, e.g. in carrot roots, or indirectly from chloroplasts, e.g. in ripening fruit. Although often found in tissues (e.g. in fruits) that do not have developmental potential, chromoplasts do not always represent a terminal developmental state, thus under some circumstances they can revert to green, photosynthetic chloroplasts, e.g. when carrot roots turn green in the light. Their formation is under developmental and genetic control, witness the ripening of green *Capsicum* fruit to a red colour, but only in genotypes that possess genes for accumulation of the carotenoid capsanthin (the major, but not the only, carotenoid pigment of *Capsicum* chromoplasts). Much effort by plant breeders goes into selecting and manipulating genes for maximising production of specific carotenoids. Apart from their nutritional value as precursors of vitamin-A, lycopene (e.g. in tomato fruit) and β-carotene (e.g. in carrot roots) have uses in food colouration. Chromoplasts retain their complement of

plastid DNA, in nucleoid areas, but the genes that govern pigment biosynthesis in fact reside in the cell nucleus.

The morphology of the pigment deposits in chromoplasts varies considerably. Droplets in the stroma are common; fibres, narrow tubules and crystalline inclusions (often so large and of such defined shape that they distort the shape of the entire plastid) also occur. If the chromoplasts have developed from chloroplasts the grana and stroma thylakoids are usually modified or may even degenerate completely.

Plate 29a Chromoplasts dominate the cytoplasm in this part of a cell from the orange rim of the corolla tube of a *Narcissus poeticus* flower. Their outlines, and especially their internal membranes, are convoluted. The clear zones within them (stars) represent crystals of β-carotene. It is now electron transparent because some at least of the β-carotene has been extracted during dehydration of the specimen, after the shape of the crystals was preserved by fixation. Numerous thylakoids undulate through the electron transparent areas. Many of the chromoplasts contain electron dense globules - plastoglobuli, which may, like the crystals, contain chromoplast pigment, but probably also contain (as in chloroplasts) plastid quinones. The upper left portion of the micrograph is occupied by part of the nucleus. It contains dense heterochromatin, and is penetrated by cytoplasmic channels (asterisks) lined by pore bearing (arrows) nuclear envelope. The cytoplasm is vacuolate (V), with mitochondria (M), rough endoplasmic reticulum, free ribosomes and lipid droplets (L). x15,000.

Plate 29b The chromoplast shown here (from a tomato fruit) is in a relatively early developmental stage. Its juvenility is shown by the presence of many small grana: later in development these disappear, leaving electron dense plastoglobuli and crystals of lycopene (lycopene is a precursor of β-carotene, which also occurs in tomato chromoplasts, but (except in certain varieties) in concentrations that are too low to give extensive crystallization).

The lycopene crystals (stars) have angular outlines, and are surrounded by membrane. They also contain undulating membranes, sometimes aggregated in electron dense stacks (e.g. see crystal in chromoplast at left hand side). In these respects lycopene crystals resemble crystals of β-carotene (see (a)). Other features shown in the chromoplasts include a round membrane-bound inclusion (arrow), aggregated electron dense material (open arrow, possibly remnants of a starch grain, or components of plastoglobuli undergoing crystallization). The stroma contains plastid ribosomes and nucleoid areas (large circle). The double envelope of the chromoplast is also visible (small circles).

Tomato fruit parenchyma cells are very large and vacuolate (V, vacuole; T, tonoplast). A microbody is included in the micrograph (asterisk). x32,000.

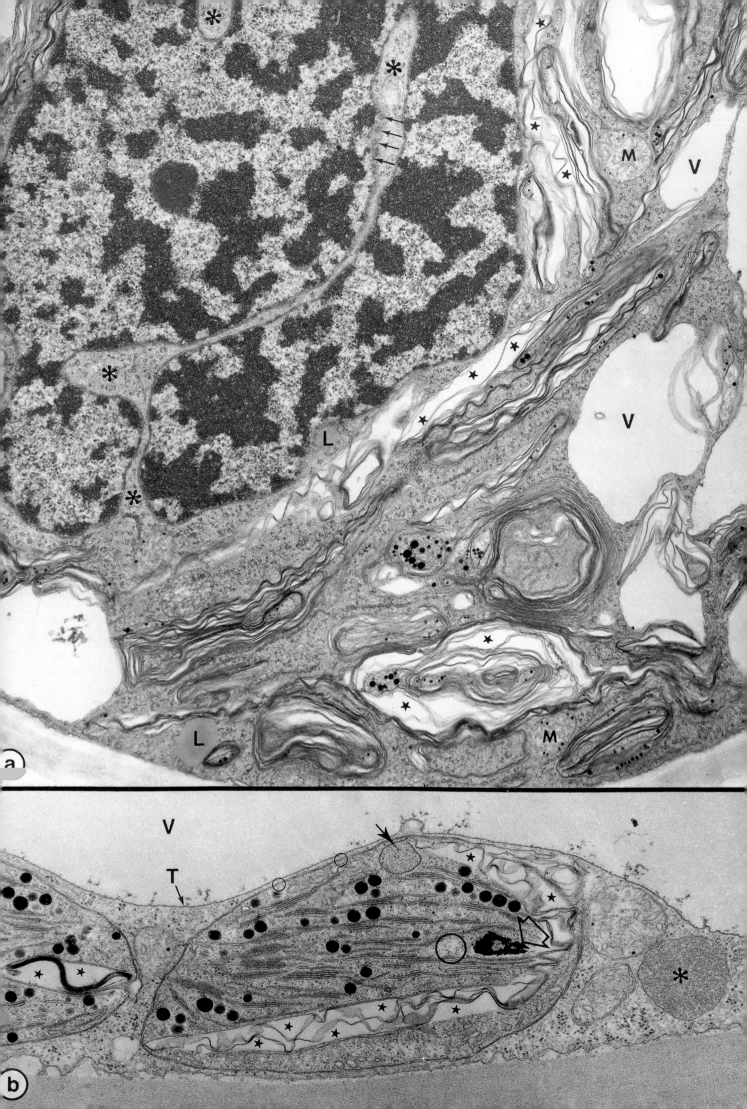

30   Microbodies

Introduction: The term *microbody* refers to a category of enzymatically-diverse compartments found in most if not all eukaryotic cells. They are structurally simple but biochemically diverse. A single membrane encloses a matrix. The feature that they have in common, wherever they occur, is that they compartmentalise (and hence protect the rest of the cell from) enzymes that are essential but have the hazardous property of producing hydrogen peroxide. Microbodies contain the enzyme catalase in large quantities (often forming a crystalline core in the matrix) to remove this very reactive, toxic compound. Catalase degrades peroxide to water and molecular oxygen.

The diversity of microbodies lies in the variety of biochemical pathways that make use of peroxide-producing oxidase enzymes. Several types of plant microbody have been characterised and two, *peroxisomes* and *glyoxysomes*, are shown here. Peroxisomes occur in photosynthetic cells. Along with chloroplasts and mitochondria, they metabolise products of the oxygenase reaction of rubisco (see Plates 23,24). Glyoxysomes are concerned with utilisation of fatty acids derived from lipid stores, e.g. in seeds. A third type contains enzymes for metabolising nitrogen-rich products of nitrogen-fixation in ureide-producing leguminous root nodules. Another occurs in senescing leaves, breaking down nitrogenous compounds. Further work may well bring other categories to light.

Plate 30a   The microbody (MB) in this section of a tobacco leaf is a peroxisome, functioning in the enzymatic processing of glycollic acid, the 2-carbon compound produced by the wasteful oxygenase reaction of rubisco. Peroxisomes usually lie in close contact with chloroplasts, as shown here. They take in the glycollate and begin to process it so that some of its carbon can be retrieved. The first step is an oxidation to produce glyoxylate. The inevitable consequence of the chemistry of this reaction is that hydrogen peroxide is liberated. However, the high content of catalase in the peroxisomes removes the toxic peroxide. Mitochondria (M) are involved in later steps of the reaction pathway and usually lie close to the peroxisome-chloroplast complex. The peroxisome shown here contains a large crystal (CY) in a diffusely granular matrix. Its single bounding membrane (arrows) is closely appressed to the outer membranes of the two adjacent chloroplasts. The latter contain grana (G), frets (F), ribosomes (CR) and an invagination of the inner envelope membrane (I).

The presence of catalase in the peroxisomes is demonstrable by cytochemical detection. Pieces of tissue are fixed with glutaraldehyde and incubated with the artificial substrate, 3,3' - diaminobenzidine, and hydrogen peroxide. A stable precipitate is deposited where catalase acts on this substrate, and is then made electron-dense by post-fixation with osmium tetroxide (see inset). Catalase activity occurs throughout the peroxisome, but is concentrated in the crystal (CY). Where the peroxisome membrane abuts the neighbouring chloroplasts (arrows), there is a stronger staining reaction than elsewhere. There may be high catalase activity in regions of close contact, where transport of molecules between chloroplast and peroxisome is presumably intense. x41,000; inset x44,000. Kindly provided by S.E. Frederick and E.H. Newcomb, reproduced by permission of Rockefeller University Press from *J. Cell Biology*, **43**, 343, 1969.

Plate 30b,c   The glyoxysome category of microbody occurs in cotyledons cells of lipid-storing seeds, e.g. the sunflower illustrated here. Glyoxysomes (MB) in these cells have an irregular shape. As in 30a, their single membrane bounds a dense, crystal-containing (CY), granular matrix. Lipid reserves are broken down during early stages of germination and seedling growth. The lipid droplets (L) in (b) possess a surface skin of lipid molecules oriented into a monomolecular layer by contact with the aqueous environment of the cytoplasm (arrows). An oxidation that generates hydrogen peroxide is an early step in preparing fatty acids derived from the lipid for dismemberment and mobilisation of their carbon. Glyoxysomes lying alongside the source of the fatty acids detoxify the peroxide. They then proceed with a set of reactions called the *glyoxylate cycle*, in which 2-carbon pieces cut sequentially from the ends of the long-chain fatty acid molecules are used to synthesise sucrose.

(b) represents a stage four days after germination. (c) represents seven days after germination, after all or most of the stored lipid has been consumed and the cotyledons have enlarged, become green and started to function as photosynthetic leaves. The microbodies (MB) in (c) are in the same cell-type as those in (b), but have become associated with the newly developed chloroplasts (C). *Either* the glyoxysome type of microbody has metamorphosed into a peroxisome type as part of the transition in cotyledon function, *or* a second population of microbodies, with peroxisomal-type functions, has appeared. The inset shows immuno-gold labelling of an intermediate stage in cotyledon development, using specific antibodies to prove that the *same* microbody contains the enzyme isocitrate lyase (characteristic of the glyoxysome stage, labelled with large gold particles) together with serine:glyoxylate transferase (characteristic of the peroxisome stage, labelled with small gold particles). Coexistence of the two enzymes shows that the original microbodies are reprogrammed rather than replaced. It is likely that the same microbody would be subjected to yet more reprogramming when the leaf starts to senesce. (b) x31,000; (c) x43,000. (b) and (c) kindly provided by P. Gruber and E. Newcomb, reproduced by permission from *Planta*, **93**, 269, 1970, inset x40,000, courtesy of D. Titus and W. Becker, reproduced by permission from *J. Cell Biol.* **101**, 1288, 1985.

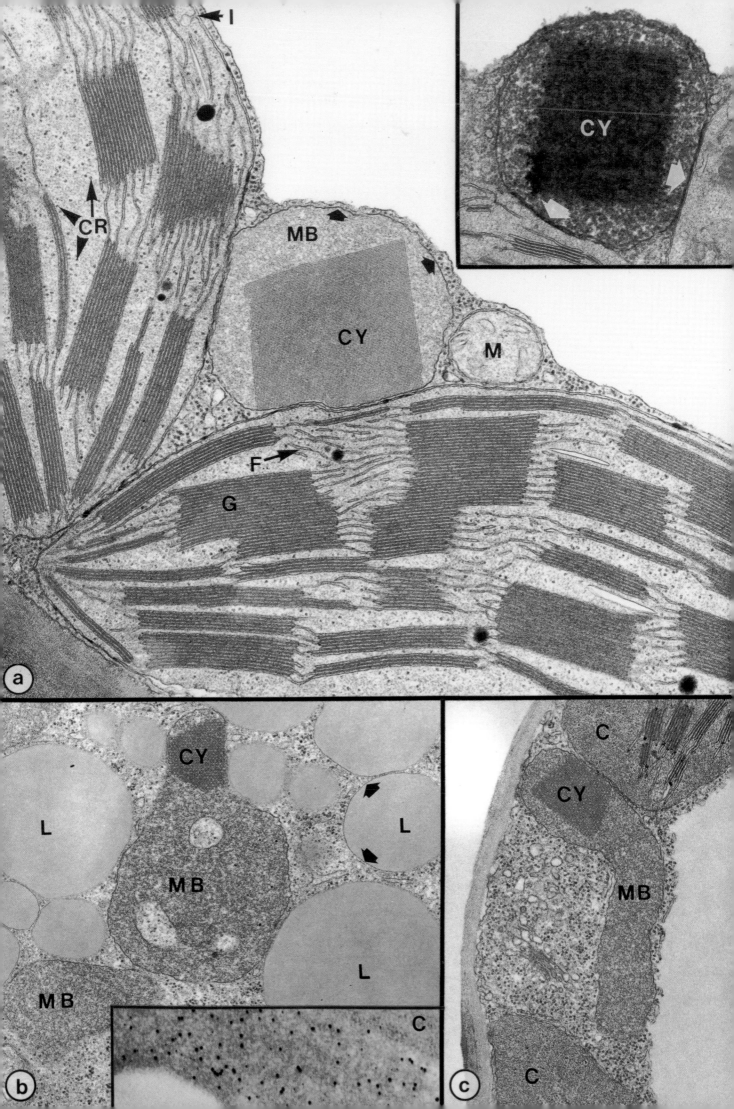

31   The Cytoskeleton: F-Actin and Microtubules

Introduction: This plate introduces two components of the cytoskeleton. *Microfilaments* (MFs) are polymers (*F-actin*) of the monomeric protein *G-actin*. *Microtubules* (MTs) are polymers of the protein *tubulin*, which occurs as dimers of α- and β-tubulin subunits. A third component of the cytoskeleton, members of the *intermediate filament* family of proteins, is more prominent and better known in animal cells, but forms of it probably occur in plant cytoplasm and in the nucleus.

Actin and tubulin both exist as pools of free proteins that are in dynamic exchange with the polymeric forms. Polymerisation requires energy; ATP to make free G-actin capable of joining the growing end of a MF and GTP (guanosine triphosphate) to allow tubulin dimers to join the growing ends of the 13 *protofilaments* that make up the wall of the MT (see diagram below). G-actin and tubulin are polarised molecules, and because they join on to the growing polymers in particular orientations, MFs and MTs are also polarised. This has great biological importance. Growth rates and rates of dissociation differ at the two ends. "Capping" molecules and molecules that initiate and assist growth exist. As a consequence, MFs and MTs can be disassembled and reassembled in different conformations, either with the same (but re-positioned) functions, or with different functions. Some of their dynamism is programmed as part of normal cell development; other aspects can be triggered externally (e.g.) by wounding, infection by a pathogen, changed light intensity or direction, or application of a hormone.

Actin and tubulin proteins do not themselves carry all of the information that is necessary to specify all of the functions of MFs and MTs. However they do carry binding sites for *actin-binding proteins* and *microtubule-associated proteins* of various sorts. These in turn confer specific roles. Some bridge MTs or MFs to each other or to other cellular structures, and are therefore important in constructing bundles or other kinds of MT or MF array. Some can undergo molecular conformational changes that allow them to "walk" along MFs or MTs. This is the molecular basis of many types of intracellular movement e.g. propulsion of *myosin* molecules along F-actin drives cytoplasmic streaming in plants and muscle contraction in animals. *Dyneins* and *kinesins* are microtubule *motor proteins*, sensitive to the molecular polarity of MTs and able to move along the protofilaments in a given direction, towards or away from the fast-growing end.

Plate 31a Rapid freezing of tissue followed by freeze-substitution and ultra-thin sectioning is a superior preparation procedure for looking at the cytoskeleton. This procedure was used here to examine a hair on the surface of a young wheat leaf. Such cells show rapid cytoplasmic streaming in life, which correlates with the presence of numerous MFs, either single (MFS) or in bundles (MFB). The hair was approximately cylindrical, and this correlates with the orientation of cortical MTs, nearly transverse to the cell's long axis. The cell, and all of the visible MTs, have been sectioned obliquely (a MT in cross section appears as a stained circle, 24nm in diameter, see diagram and Plate 32a). Microtubule-associated filaments are also visible in some areas, with minute bridges to the adjacent MTs (arrowhead). These filaments are of unknown nature, but could be F-actin. Scattered polyribosomes and vesicles are present. One vesicle has been sectioned close to its surface, revealing a clathrin coat (arrow) (see Plate 15). A clathrin lattice is also seen in face view on the plasma membrane (open arrow). Kindly provided by S. Tiwari; x70,000.

Plate 31b This root tip cell of *Phleum pratense* is sectioned parallel and close to one face of the cell to show a dense array of cortical MTs (about 15 per μm) underlying the plasma membrane. Portions of cell wall and obliquely-sectioned plasma membrane appear at the left and right hand sides of the picture. The micrograph includes microtubule-associated filaments (arrowheads), endoplasmic reticulum cisternae lined up along MTs just under the plasma membrane (ER); coated vesicles (arrows) and non-coated vesicles among the MTs). x64,000.

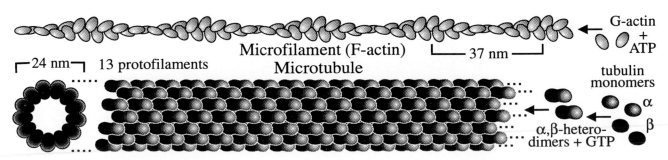

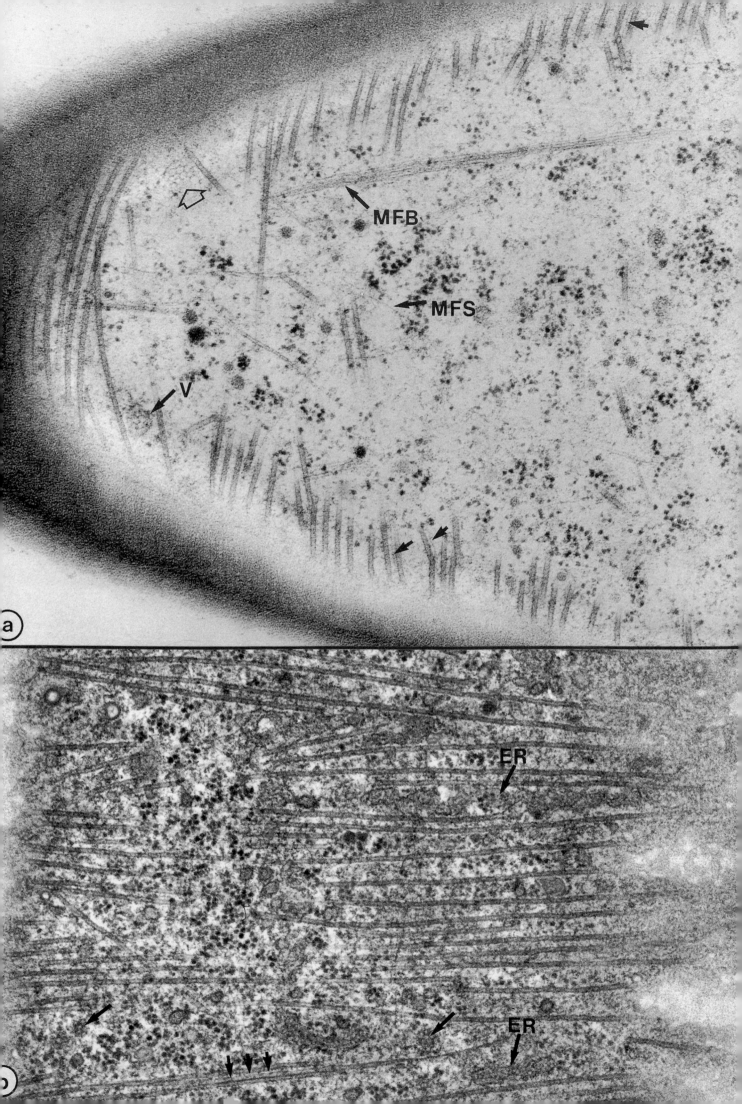

32 Cortical Microtubules

Introduction: Cortical microtubules (MTs) in plants affect cell shape in two different ways, both illustrated here. Most plant cells have walls and their cortical MTs shape cells indirectly by regulating the orientation of cellulose wall microfibril deposition. In the few wall-less cells in plants, cortical MTs have a direct skeletal role.

Plate 32a This is an example of a wall-less cell, the motile male sperm of a fern, *Blechnum nudum*. The cell body is narrow, coiled, and gyrates as it swims, propelled by the beating motion of about 50 flagella at the cell apex. This small part of the cell shows two classes of sub-plasma membrane MTs. A thin film of cytoplasm between an extremely heterochromatic nucleus (N) and the plasma membrane contains a close-packed array of MTs, each seen in transverse section as a circular profile 24nm in diameter. There being no cell wall to provide structural rigidity, these MTs have a truly cyto-*skeletal* role. They shape the cell body (as can be demonstrated by depolymerising them with an anti-microtubule drug), and the array accommodates flexing during swimming.

The micrograph also shows cross sections of some of the flagella, each of which protrudes from a basal body (e.g. arrowhead) anchored in the peripheral cytoplasm at the apical end of the cell. Each flagellum is a cylindrical outgrowth of plasma membrane (arrows), supported by a ring of nine sub-membrane MT-doublets and an axial pair of MTs. This conformation is remarkably standard throughout plant, animal and protistan kingdoms. Flagellar beating results from force generation by the microtubule motor *dynein*, which bridges adjacent MT doublets and causes them to move slightly relative to one another. The outcome is that the flagellum bends. A complex arrangement of inter-microtubule bridges is seen in the basal body, probably concerned with transducing signals from the cell body to regulate flagellar activity. The fibrillar object (BP) is a *blepharoplast*. It generates flagellar basal bodies during sperm cell development. Kindly supplied by C. Busby. x62,000.

Plate 32b This grazing section of the cell surface includes cell wall (CW), plasma membrane (PM) and the cortical cytoplasm (CP). It complements Plate 31b by demonstrating correspondence between the orientation of cortical MTs and that of cellulose microfibrils in the cell wall (analysed further in Plates 33 and 34a,b). Wall microfibrils (MF) run horizontally across the lower part of the micrograph (lower arrow), with a similarly-oriented microtubule (1) nearby. Above this a band of microtubules (2) runs at a slight angle to (1), and again the adjacent wall microfibrils parallel this orientation (arrow at right centre). A further band of microtubules (3) runs at an angle across the main group. The irregular nature of the cell surface generates complex images of bumps and hollows in the plasma membrane (e.g. circled areas) and of gaps between the wall and the plasma membrane (irregular electron-transparent areas). Vesicles (V), possibly of Golgi origin, lie among the cortical MTs. Spinach root tip (*Beta vulgaris*) x50,000.

Plate 32c This plane of section is at right angles to that in 32b, in a transverse section of a root tip cell with a very thin cell wall and numerous plasmodesmata. It shows transversely-oriented cortical MTs in longitudinal profile, close to the plasma membrane. While most of the MTs gradually taper out of the section (and would continue in an adjacent section), one is seen in its entirety, nearly 2μm long. Arrowheads mark its abrupt ends, which are quite distinct from MTs that pass out of the section. Such views, and reconstructions based on tracking MTs through serial sections (diagram below), show that cortical MT arrays consist of discrete, overlapping MTs, rather than very long MTs that wrap right round the cell. The reconstruction of course represents the array as it was at an instant in time. In life the arrays are dynamic. Individual MTs are continually shortening, growing, shifting or being replaced, although the general orientation persists. *Azolla* root cell, x50,000; reconstruction kindly provided by A. Hardham.

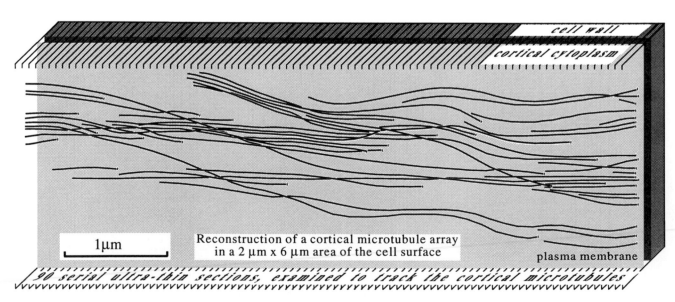

Reconstruction of a cortical microtubule array in a 2 μm x 6 μm area of the cell surface

1μm

cell wall

cortical cytoplasm

plasma membrane

90 serial ultra-thin sections, examined to track the cortical microtubules

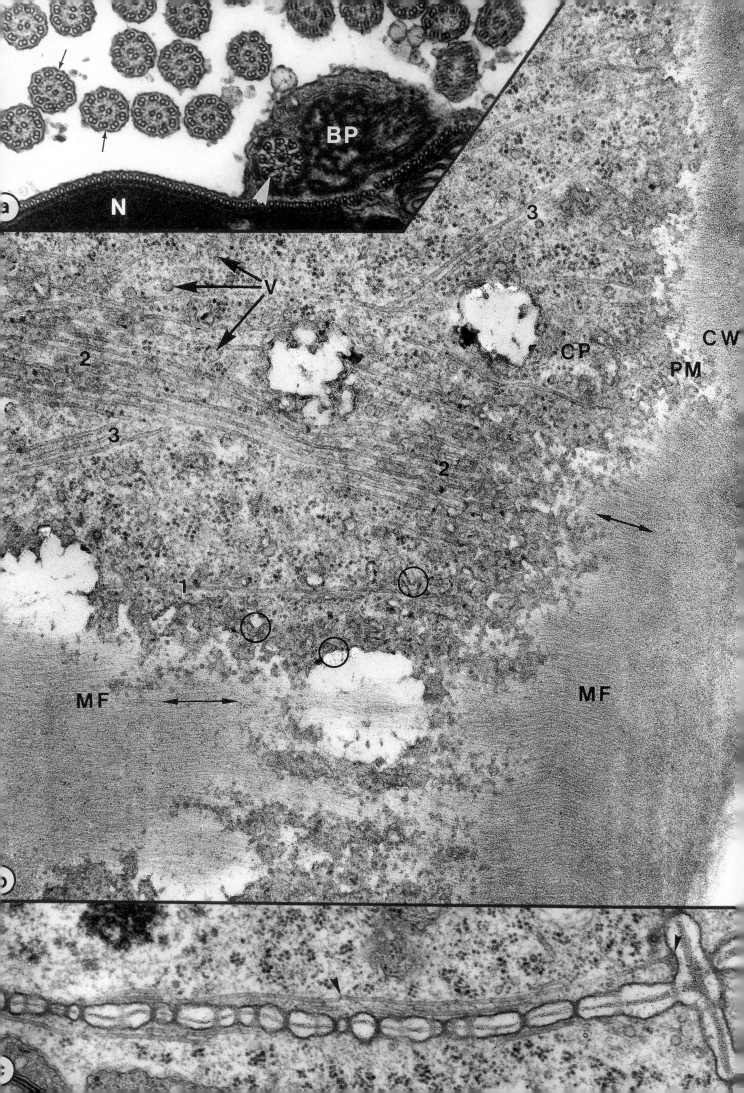

33   Cell Wall Microfibril Synthesis

Introduction: Synthesis of cellulose involves joining glucose units into long chains by $\beta(1\rightarrow4)$ links, and these chains into bundles - *microfibrils* - at the cell surface. The molecular machinery consists of arrays of integral membrane proteins, best seen on the protoplasmic face (PF) of freeze-fractured plasma membranes. The hexagonal rosettes illustrated in this plate and in the diagram (below) occur in higher plants. Conformations found in other plant groups include single particles, various linear complexes of particles, and extensive planar lattices of hexagons. Sometimes bulges occur in the corresponding extraplasmic face (EF), at the terminus of the emerging microfibril. There is much circumstantial evidence that these complexes make microfibrils. Thus their numerical density correlates with the rate and timing of cellulose synthesis. They occur in known sites of wall synthesis. The rosettes are synthesised in the endoplasmic reticulum and inserted into its membrane for delivery to the plasma membrane *via* secretory Golgi vesicles. They are thought to be active for short periods of time only, perhaps as little as 10 minutes, before degradation and replacement by new rosettes. Inhibitors of protein synthesis reduce their numbers *and* microfibril synthesis. Cortical microtubules impart directionality to deposition of new microfibrils (Plate 32), perhaps by guiding movement of cellulose synthesising complexes in the plane of the plasma membrane (see diagram).

Plate 33a This freeze fracture face is viewed as if from the inside of a cell looking out towards the wall. The fracture plane has passed at different levels across the cell surface, exposing wall microfibrils to view at the extreme right (F) and membrane particles on the EF face on the left, having passed along the hydrophobic interior of the plasma membrane, between the lipid bilayer leaflets. In life this face was tightly appressed to the wall in a turgid cell, and after rapid freezing it still carries impressions of numerous wall microfibrils (F) that were pressed against it at the inner face of the wall. In the main area to the right and top of the micrograph the fracture has exposed a layer of cytoplasm just inside the plasma membrane. Microtubules (MT) accompanied by parallel microfilament bundles (MF) are revealed. Note that the general alignment of microfibrils in the wall under the EF face parallels the orientation of the microtubules. Root cell of *Trianea bogotensis* (syn *Limnobium stoloniferum*), x24,000.

Plate 33b Here the fracture face exposes the PF face of the plasma membrane, viewed as if from the outside of the cell with the EF layer removed. Numerous membrane particles lie either singly, or in occasional unordered clusters of 2-3 particles, or in well defined rosettes of 6 closely associated particles (examples arrowed). Root xylem element of cress, *Lepidium sativum*, x200,000. Both micrographs kindly provided by W. Herth.

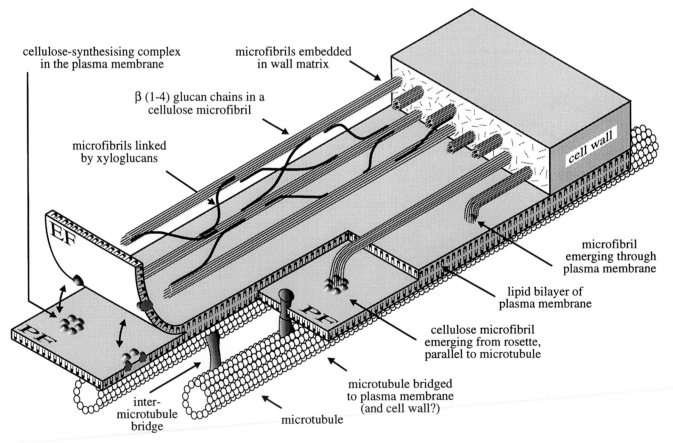

cellulose-synthesising complex in the plasma membrane

microfibrils embedded in wall matrix

β (1-4) glucan chains in a cellulose microfibril

microfibrils linked by xyloglucans

cell wall

EF

PF

PF

inter-microtubule bridge

microtubule

microtubule bridged to plasma membrane (and cell wall?)

cellulose microfibril emerging from rosette, parallel to microtubule

lipid bilayer of plasma membrane

microfibril emerging through plasma membrane

Relationships of cellulose-synthesising complexes (rosette form), wall microfibrils, plasma membrane and microtubules.

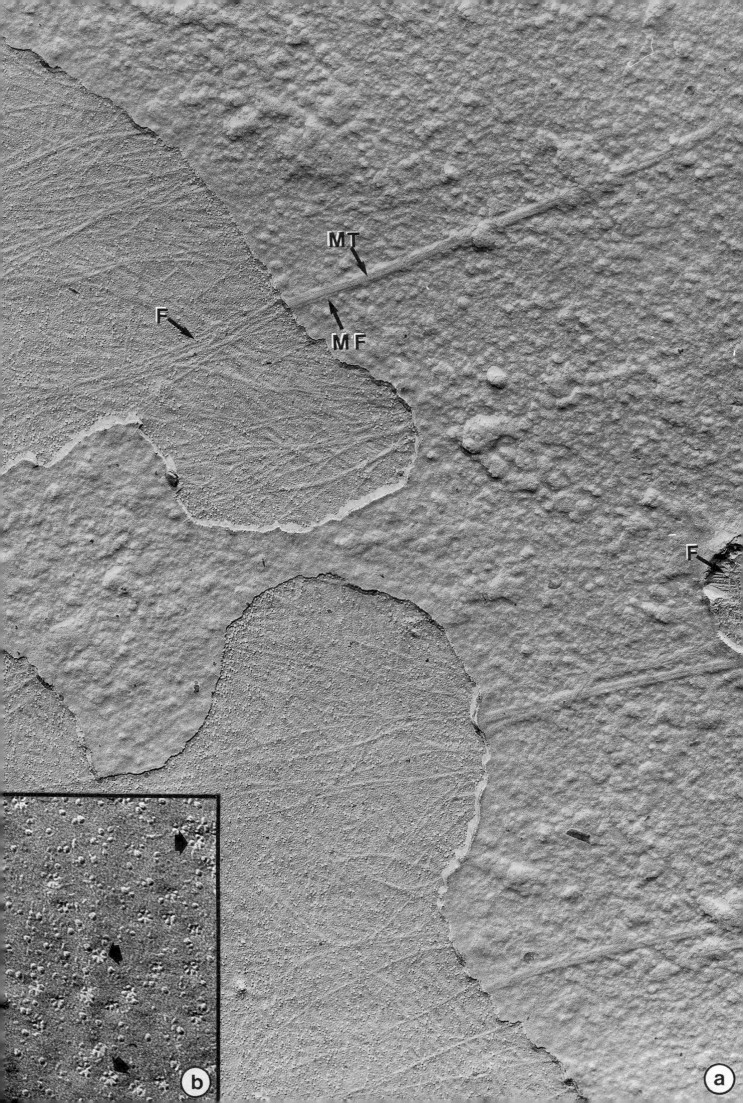

34 Microtubules in Interphase and Cell Division

Plate 34a,b These cells in the freshwater green alga *Nitella* illustrate the morphogenetic role of cortical microtubules (MTs) that was introduced in Plates 31-33. (a) is the apex of a plant, showing a cylindrical internode cell and rosettes of single-celled branches. (b) is a comparable apex from a plant that was treated with the anti-MT herbicide oryzalin. Cells that normally grow as cylinders have developed an almost spherical shape. In (a) the cylindrical shape is generated because cellulose microfibrils are deposited circumferentially in the cell wall, transverse to the long axis, allowing the cells to elongate much more easily than they can increase in girth. Oriented deposition of cellulose is guided by cortical MTs (Plate 33). Oryzalin removes the cortical MTs and the growing cells then have no control over cellulose orientation. With randomly oriented cellulose, the mechanical constraint on the geometry of growth is lost and the cells expand in all directions. Cells that had *already* grown as elongated cylinders (lower left and right of (b)) at the time of the treatment are unaffected. (a) and (b) x20. (b) kindly provided by G. Wasteneys.

Plate 34c MTs in a young cell from the cortex of a wheat root tip. This hexagonal cell, with flat upper and lower surfaces, was isolated by enzymatic digestion of its cell walls and its cortical MTs were revealed by the fluorescent antibody technique using anti-tubulin. Many MTs lie on the side (longitudinal) faces of the cell, predominantly transverse to the axis of elongation of the cell (and the root). There are very few MTs on the transverse faces of the cell.

Plate 34d,e MTs can also be visualised in sections by immunofluorescence microscopy. The example here is a section of a root meristem. It is included to introduce the four categories of MT array that occur in cells in different stages of the cell division cycle and in cells that are starting to elongate after division. (d) shows the arrangement of cells in the apical meristem of an *Azolla* root, seen in a median longitudinal section and (e) shows a similar section, stained with anti-tubulin. Transversely-oriented *cortical MTs* occur against the side walls of cells that are not engaged in division (arrows). *Preprophase bands* of MTs occur in cells that are preparing to divide (arrowheads). Some cells are making *mitotic spindles* (open arrows). The apical cell of the root is undergoing cytokinesis and contains a *phragmoplast*. See also the diagram of MT arrays during the cell division cycle (below), and Plate 37. (d,e) x1,450.

Plate 34f,g,h Sections of onion root tip cells triple-stained with anti-tubulin (yellow-green), a DNA stain (blue) and calcofluor (cell walls, cyan) to show (f) cortical MTs lying close to the cell surface, (g) a preprophase band with initial spindle development at the poles of the nucleus, (h) a cell late in division, with a phragmoplast between two separated groups of chromosomes. A phragmoplast is a dense array of MTs in which a new cell wall develops at the end of mitosis, first at the centre line and later growing out to the edges of the parent cell. The new wall fuses with the parental walls where the preprophase band lay before mitosis started. Plate 40 gives more details of preprophase bands. (f,g,h) x1,800.

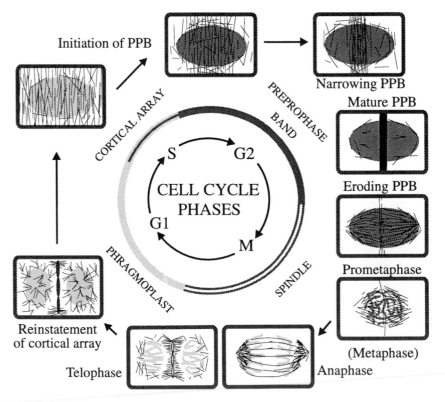

Diagram of MT arrays in relation to cell cycle phases: DNA duplication occurs in a period that lasts several hours, known as the "S-phase" of the cell cycle. Mitosis ("M-phase") commences after a "gap" ("G2 phase") . The "G1-phase" follows mitosis, to complete the cycle. Changes in the MT system are geared to these events. Cortical arrays are found during G1- and S-phases. The preprophase band begins to form at the end of the S-phase and is fully condensed just before the nuclear envelope breaks at the end of prophase of mitosis. It is then supplanted by the MTs of the mitotic spindle. They in turn give way to the phragmoplast MTs. Finally the cortical array is reinstated. Plate 37 illustrates these events.

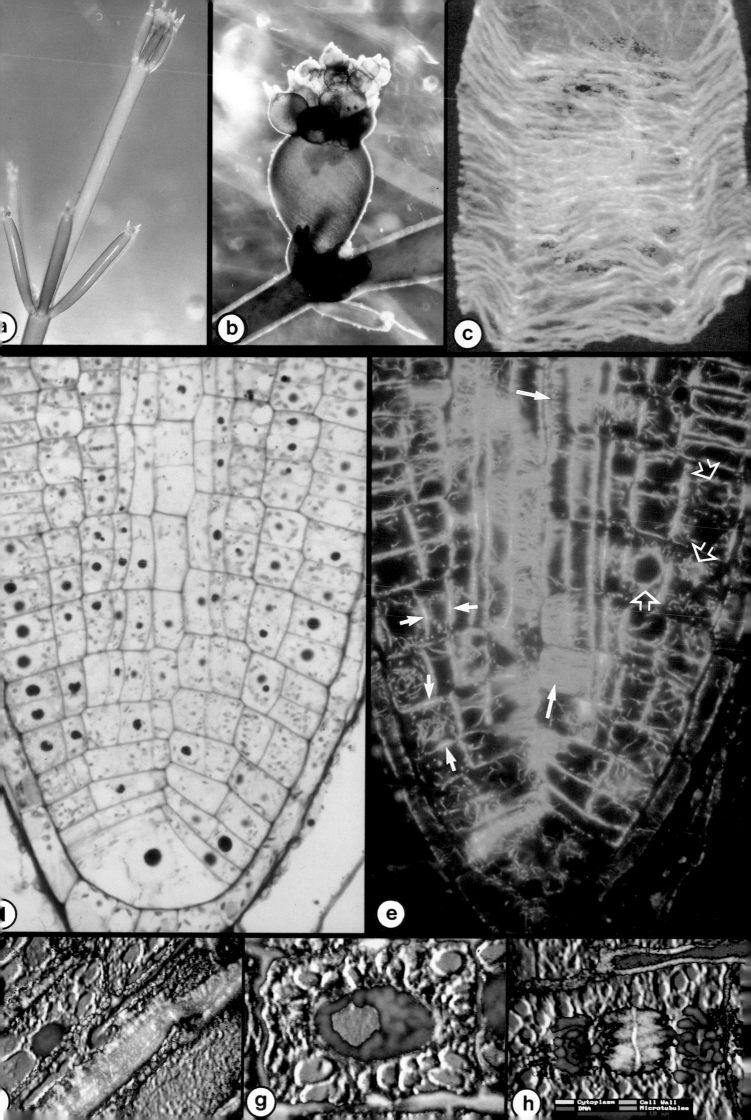

35   Microfilaments of F-Actin

Introduction: *Cytoplasmic streaming* is a vital process in plant cells. It distributes and stirs the cytoplasm, moving sub-cellular components along strands that ramify into all regions of the cell. If left un-stirred, movement of enzymes and substrates would depend solely on the physical process of diffusion. This would be inefficient over long distances, such as from one end of an elongated plant cell to the other, perhaps several hundreds of micrometres away. Cytoplasmic streaming distributes substrates and metabolites from sites of production in the cell (or entry into the cell) to sites of utilization (or exit). It is especially important where the cytoplasm is in the form of thin films or strands in large, vacuolated cells, where the paths for movement are long, tortuous and of narrow cross-sectional area. In most cases streaming occurs along microfilaments (MFs) of F-actin, propelled by acto-myosin interactions (see Plate 31).

The pictures in this plate are reconstructions of F-actin MFs arrays obtained by confocal laser scanning fluorescence microscopy. Successive focal planes in the specimens were recorded digitally and were then summed to give single projections with extended depth-of-focus. F-actin is shown in green-yellow; nuclear DNA in blue. Plates 31 and 57 show F-actin in electron micrographs.

One great advantage of fluorescence microscopy is that by making objects emit light (fluorescence), it can render them visible even if they are smaller than the resolution limit of the microscope. However, for interpreting the images, it also needs to be noted that very small objects, like MFs or MTs, inevitably appear larger than their true size. This is because microscope lenses cannot deliver better performance than their *diffraction limit*. For glass lenses, light emitted from a theoretical point source can never be imaged smaller than about 0.2 µm diameter, at best. Single MFs or MTs are about ten times thinner than this diffraction limit, so they appear, not at their true size, but inflated about ten-fold in thickness. It is therefore not possible to say (for example) how many strands of F-actin (or MTs) are present in narrow fluorescent bundles. Two neighbouring filaments cannot be resolved if their individual inflated images overlap. The higher resolution of the electron microscope remains necessary for such matters.

Plate 35a These cells in a longitudinal slice of the zone of cell elongation of a wheat root tip were labelled with an antibody that recognises actin specifically. Abundant MFs of F-actin pass into all parts of the cytoplasm, often looping around the nucleus. Their thickness varies considerably, presumably because the bundles contain variable numbers of strands of F-actin. Cell walls are not stained, but the single nucleus, and the conformation of the MFs, indicate the boundary of each cell. MFs do not pass from one cell to another. Small cells near the outside of the root (top) are underlain by more elongated cells (bottom). In general, the longer the cell, the more important cytoplasmic streaming is in its physiology, and the greater the length and prominence of its MF system. x1,900.

Plate 35b,c,d The toxic peptide phalloidin, found in a group of fungi, is very poisonous because of its high affinity for F-actin, a property exploited here for the purpose of localising MFs. Fluorescent-labelled phalloidin was microinjected into the cytoplasm of a living *Nitella* internode cell, like the one shown in Plate 34a.

These huge cells display the fastest of all examples of cytoplasmic streaming, at rates greater than 50µm per second. A static outer cytoplasmic zone contains files of chloroplasts (here seen as unstained ellipsoidal bodies). Each file has several filaments of F-actin running along its innermost face, at the outer boundary of the streaming *endoplasm*. The stream passes along one half of the cylindrical cell and back along the other. The sides of the two opposing streams are separated by a stagnant *neutral line*, shown along the centre of (b). Several nuclei (blue) are visible in the neutral line. Often the files of chloroplasts and the associated F-actin lie in a helical arrangement, as in (c) (long axis of the cell is left-to-right). MF bundles occasionally pass from one file of chloroplasts to another. They curve around the ends of the cells like an endless belt (d). The actin does not itself move, but directs the streaming cytoplasm along the length of the cell, across its end, back along the other side of the cylinder and across the opposite end.

The highly organized actin cytoskeleton of *Nitella* (b,c,d) contrasts with the irregular MFs in wheat cells (a). However, both systems possess an underlying molecular regularity. As introduced in Plate 31, G-actin molecules are "polar" (one end different from the other) and when they polymerise to make F-actin the resulting filaments are also polarised. They grow predominantly in one direction and drive cytoplasmic streaming in the same direction. In *Nitella* all of the filaments in the "endless belt" share the same polarity and combine to move the streaming cytoplasm unidirectionally. In wheat the filaments are irregularly arranged and even those that lie close to one another can have opposite polarity, giving rise to very complex patterns of streaming. Thin strands of cytoplasm that contain oppositely-directed flows next to one another can often be seen. Propulsion occurs when myosin molecules on the surface of cell membranes contact F-actin and ATP is hydrolysed to generate a *power stroke*, i.e. a conformational change in the myosin, moving it (and the cell component to which it is attached) towards the *plus*, or fast-growing, end of the polarised filament. In *Nitella* the multiplicity of unidirectional power strokes throughout the MFs gives enough momentum to sweep the total endoplasm around the cell. (b) x920; (c) x1,600; (d) x1,000. Micrographs (b), (c) and (d) kindly provided by G. Wasteneys.

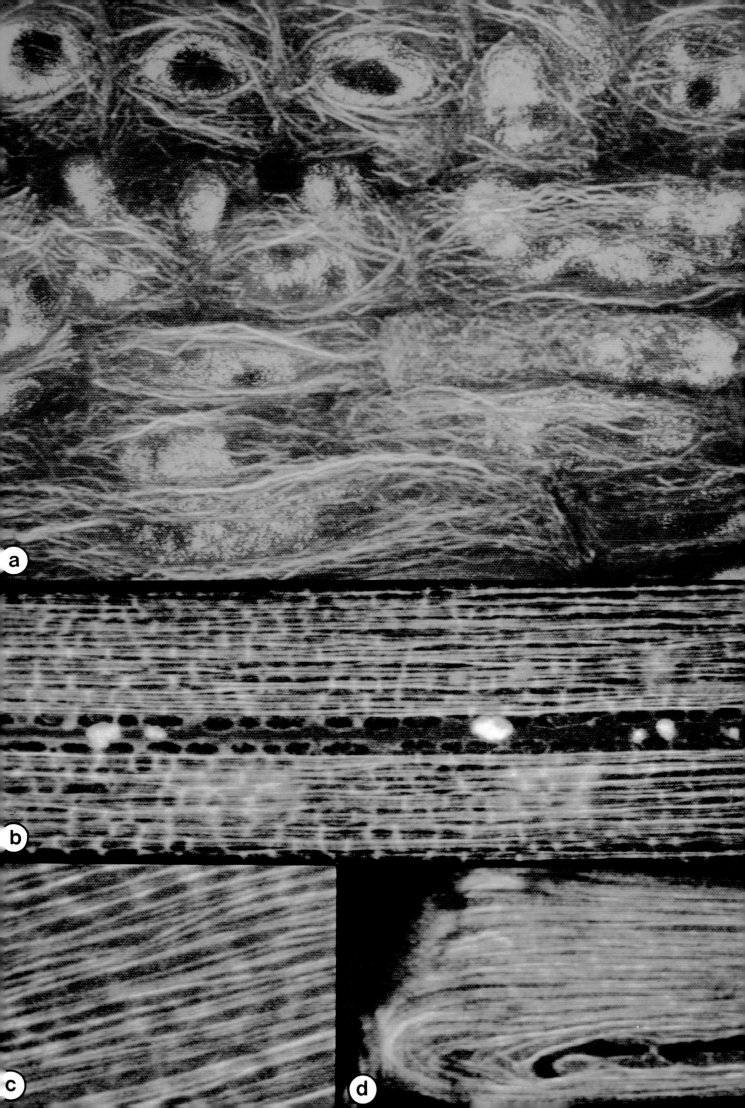

36   Endoplasmic Reticulum

Introduction: This plate supplements earlier micrographs that illustrate roles of the endoplasmic reticulum (ER) in biosynthesis (Plates 7,8). Another role of the ER is to influence local events in the cytoplasm by taking up or releasing Ca^{2+} ions, to which many processes, especially those based on the cytoskeleton, are very sensitive. In so doing the ER occupies strategic locations and undergoes dramatic changes in distribution, for example during the cell division cycle.

The distribution of the ER is revealed in (a)-(p) by immunofluorescence microscopy, using an antibody that locates the *ER retention signal sequence* of amino acids, a distinguishing feature of proteins that reside in the ER (see Plate 7 for further details). Nuclear DNA was stained separately and is shown in blue. The plane of cell division is from left to right in all cases (see also plates dealing with mitosis (37-39, 41-44). In (q,r) a fluorescent dye, $DIOC_6$, is used to visualise cortical ER in living cells. All of the images are confocal scanning micrographs. (a-p) show optical sections through isolated cells from wheat root tips; (q and r) are computed stereo images.

Plate 36a shows a cell whose condensed chromatin indicates that it was preparing to divide. Its ER ramifies throughout the cytoplasm. The nuclear envelope also reacts with the antibody, reinforcing the view that it is a special form of ER, continuous with the cytoplasmic system (Plate 5). Judging by the state of the chromatin this cell would have had a preprophase band of microtubules (see Plates 37,40), however there is no distinctive arrangement of ER at the site of the band (left and right of the mid zone of the cell). x1,800.

Plate 36b This cell is in late prophase of mitosis, with condensed chromosomes. ER cisternae have now aggregated conspicuously at the future spindle poles, outside the nucleus (equivalent to Plates 38a-c, 39a). This is where many spindle microtubules are assembling (Plate 37e,f). Microtubule formation is known to be sensitive to the concentration of Ca^{2+} ions and it may be that the ER at the spindle pole regions helps to create favourable local conditions for polymerization of tubulin. x2,000.

Plate 36c The nuclear envelope has just broken down at the end of prophase (equivalent to Plate 39d). ER persists at the nuclear poles and is beginning to penetrate among the chromosomes (arrows). x2,400.

Plate 36d,e,f,g,h *Prometaphase-metaphase stages.* ER strands now penetrate from the polar aggregations among the chromosomes. By the time the chromosomes have become oriented into a metaphase plate (g,h), ER cisternae lie between nearly all of the trailing arms of the chromosomes, close to the kinetochore bundles of microtubules (see Plates 37i-k, 42). Once again a role for the ER in microtubule assembly/disassembly is indicated. (d) x2,000; (e) x1,400; (f) x1,000.

Plate 36i,j *Anaphase stages.* The ER tends to withdraw from among the chromosomes as the daughter chromatids separate during anaphase. This is when the bundles of kinetochore microtubules shorten towards the poles (see Plate 37l-n). (g) x1,500; (h) x1,200; (i) x1,800.

Plate 36k-n *Telophase stages.* Caps of aggregated ER remain at the poles in telophase and some even becomes trapped in the daughter nuclei when their nuclear envelopes re-form. At first there is not much ER in the zone between the new nuclei, where the phragmoplast arises (see Plate 37n,o). However ER soon becomes an integral part of the forming cell plate (see also Plate 44). Cisternae of ER invest and penetrate the new wall, and the whole phragmoplast becomes shrouded in other cisternae, especially at its outwardly-growing margin. (j) 1,600; (k) x1,800; (l) x1,600; (m) x1,800; (n) x2,200.

Plate 36o,p *Cytokinesis.* The cell plate eventually reaches the parental walls; the polar aggregates of ER disperse; the DNA in the daughter nuclei decondenses and the ER returns to its starting configuration, as in (a). (o) x1,900; (p) x1,600.

Plate 36q,r Carbocyanine dyes accumulate in membranes in living cells when there is an electrical potential difference from one side of the membrane to the other. 3,3'-dihexyloxacarbocyanine iodide ($DIOC_6$) is especially good for revealing ER membranes in many kinds of plant cells, although it can also stain the tonoplast, plasma membrane (see Plate 16) and mitochondria. The dye molecules line up in fluorescent side-by-side arrays in the membrane. These stereo images, of large living cells in onion bulb scale leaf epidermis, were obtained by computer addition of many successive optical sections, one complete set in red and another in green. The "red" and the "green" sets are progressively tilted relative to each other as the successive optical sections are added, thus building depth information into the final superimposed red-green image. The stereo effect in these images must be observed by putting a red filter (e.g. cellophane) in front of the left eye and a green filter in front of the right eye. The two pictures have been printed with opposite stereo effects in order to accentuate depth perception. The cells in (q) appear to recede into the page, while (r) appears to bulge upwards (or the opposite effect if the red-green filters are reversed).

The low magnification view (q) shows trans-vacuolar strands, nuclei and faintly fluorescing images of ER at the cell surface. (r) shows cortical ER at higher magnification, in the form of flat plaques lying close to the plasma membrane (which is not visible), interconnected by fine tubules (see also Plates 8, 16). Narrow tubules of ER also occur individually and in bundles passing along the trans-vacuolar strands. However, these are not well resolved because they were in continuous rapid motion in the living cells and were therefore hard to image using confocal microscopy. (g) x360; (r) x900.

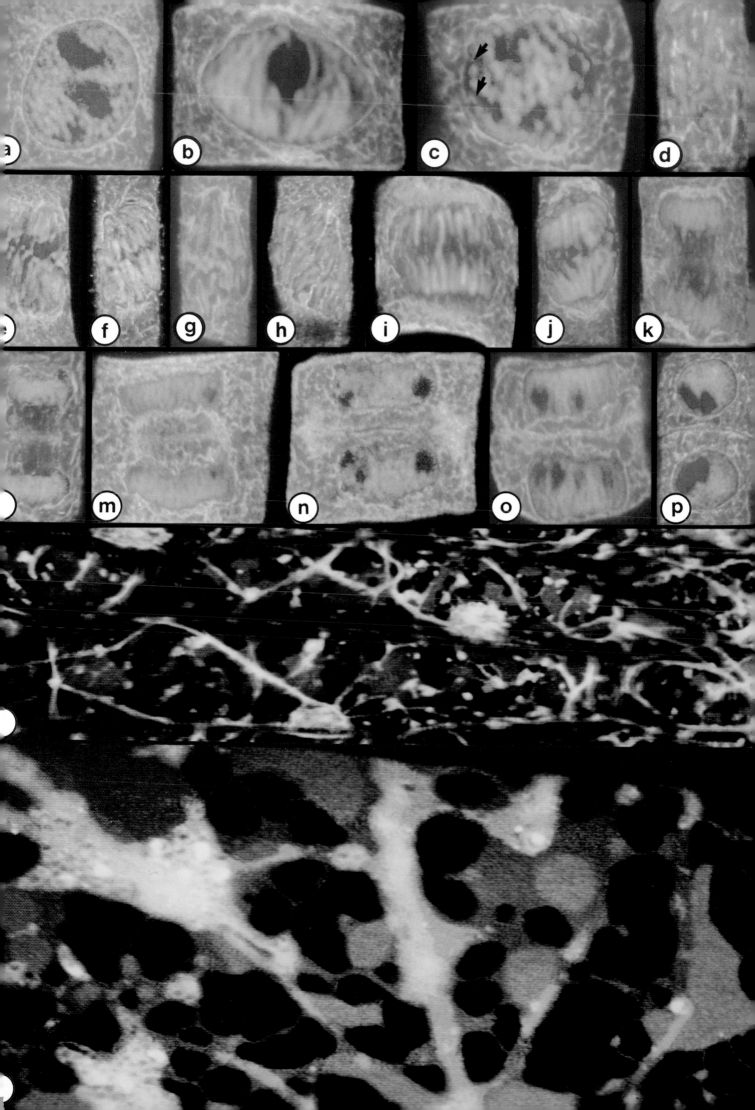

37   Microtubules during the Cell Division Cycle

Introduction: As diagrammed in Plate 34, the cell division cycle can be considered as a sequence of processes: first a period of growth to attain sufficient mass to allow division to occur (the G1 phase), then a period of DNA synthesis during which the genetic information in the cell nucleus is duplicated (the S-phase), then a gap before mitosis starts (the G2-phase), and finally mitosis (the M-phase). The whole sequence typically lasts about 18-24 hours in root tip cells, with S-phase occupying 3-4 hours and mitosis 2-3 hours. Mitosis is itself a sequence of events: prophase, prometaphase, metaphase, anaphase and telophase.

This plate shows confocal scanning micrographs of dividing cells from wheat root tips, with microtubules (MTs) stained using anti-tubulin antibody. Cell nuclei are shown in blue. Stages of the cell division cycle from the end of S- through G2-, M- and early G1-phases are included. The starting condition for the cycle is typified by Plate 34c, another wheat root tip cell that was processed in the same way. The sizes and shapes of the cells vary somewhat, but in all cases the plane of division is left to right.

Plate 37a-d *Preprophase band development.* The first sign in the MT population that a cell is starting to prepare for division is that the cortical MTs begin to aggregate. This takes place progressively throughout G2, ultimately creating a girdle of MTs called the preprophase band (PPB) at the future site of division, i.e. the site where the new cell wall that will partition the parent cell at the end of cytokinesis will fuse with the parental cell wall. (a) shows the earliest stage of PPB formation and (d) shows its culmination. The process begins as soon as the S-phase ends, i.e. when the cell has completed DNA synthesis in its nucleus, it begins to prepare for the cytoplasmic event, cytokinesis, that will complete the division process. (b) and (c) are intermediate stages of PPB formation, seen in cells with different shapes. It is not known whether the PPB consists of the same set of MTs that constituted the preceding cortical array, or whether it is a new set. However, as soon as aggregation starts, many MTs appear at the surface of the nucleus and this may be an indication that a new population is being made. By the time the PPB is mature (d) its MTs are so tightly packed that individuals cannot be picked out by fluorescence microscopy. The position of the mature PPB predicts where the cell will divide. See Plate 40 for an electron micrograph of this stage and further discussion of PPBs. (a) x2,400; (b) x1,400; (c) 2,200; (d) x2,800.

Plate 37e-i *Prophase* These views cover the transition between the maturation of the PPB and the breakdown of the nuclear envelope at the end of prophase. The mitotic spindle starts to form outside the nucleus while the chromosomes complete their condensation. At first the spindle MTs are in tufts, not precisely aligned (e) but later they lie in the pole-to-pole axis (f-h). The PPB begins to erode away as the spindle MTs become more crowded, though there is no indication that its MTs are moving into the spindle. Eventually the nuclear envelope breaks and the spindle MTs then have access to the mass of chromosomes, where additional MTs also form (i). The last remnants of the PPB are seen in (i). x2,300; (f) x2,600; (g) x1,800; (h) x2,050; (i) x2,400.

Plate 37j-m *Prometaphase-anaphase.* Bundles of MTs attach to the kinetochore regions of the chromosomes during prometaphase (j). Pole-to-pole oscillations during prometaphase generate the metaphase plate of chromosomes (k), in which all of the kinetochores lie in a plane, with bundles of kinetochore MT extending away towards both poles. The DNA was duplicated previously (in the S-phase) and by now is in the form of two chromatids in each chromosome. These now separate and move apart during anaphase (l,m). The brightly fluorescent regions where bundles of kinetochore MTs attach to the daughter chromatids can be seen to approach the poles as the bundles progressively shorten (l), until all of the kinetochores are close together at the poles (m). (j) x1,950; (k) x2,400; (l) x2,000; (m) x2,800.

Plate 37n-q *Telophase and phragmoplast development.* MT bundles that extend from each pole to the equator become visible during anaphase (l,m). They were present earlier, but were obscured by the chromosomes. Now they become progressively augmented in the mid zone by new MTs representing the first stages of phragmoplast development (n, o). The cell plate begins to show up as a narrow non-fluorescent zone at the mid-line of the phragmoplast. As well as its numerous MTs, the phragmoplast apparatus contains a motor protein that is thought to move vesicles of cell wall precursor along MTs into the new cell plate, and abundant cisternae of endoplasmic reticulum (see Plate 36k-o). It grows outwards to the cell periphery (p,q), directed to the precise site where the PPB developed in G2. Meanwhile the chromosomes decondense and the daughter nuclei become foci for formation of new MTs, prior to reinstatement of cortical arrays in the daughter cells (p,q). (n) x2,500; (o) x2,150; (p) x2,400; (q) x1,800.

Plate 37r-v *Completion of cytokinesis and reinstatement of cortical arrays of MTs.* MTs disappear from the central region of the phragmoplast as it grows outwards (r) and eventually, when cytokinesis is almost complete (s, t, u), only a few fragments remain at the division site. More and more new MTs appear in the cytoplasm around the daughter nuclei. These are then supplanted by MTs at the cell surface, where progressively denser and more oriented new cortical arrays arise. (u) shows the first stages of cortical alignment and (v) shows a later stage, with nearly all MTs aligned transversely on one (but not all) of the faces of each daughter cell. (r) x1,300; (s) x2,500; (t) x1,800; (u) x1,550; (v) x2,600.

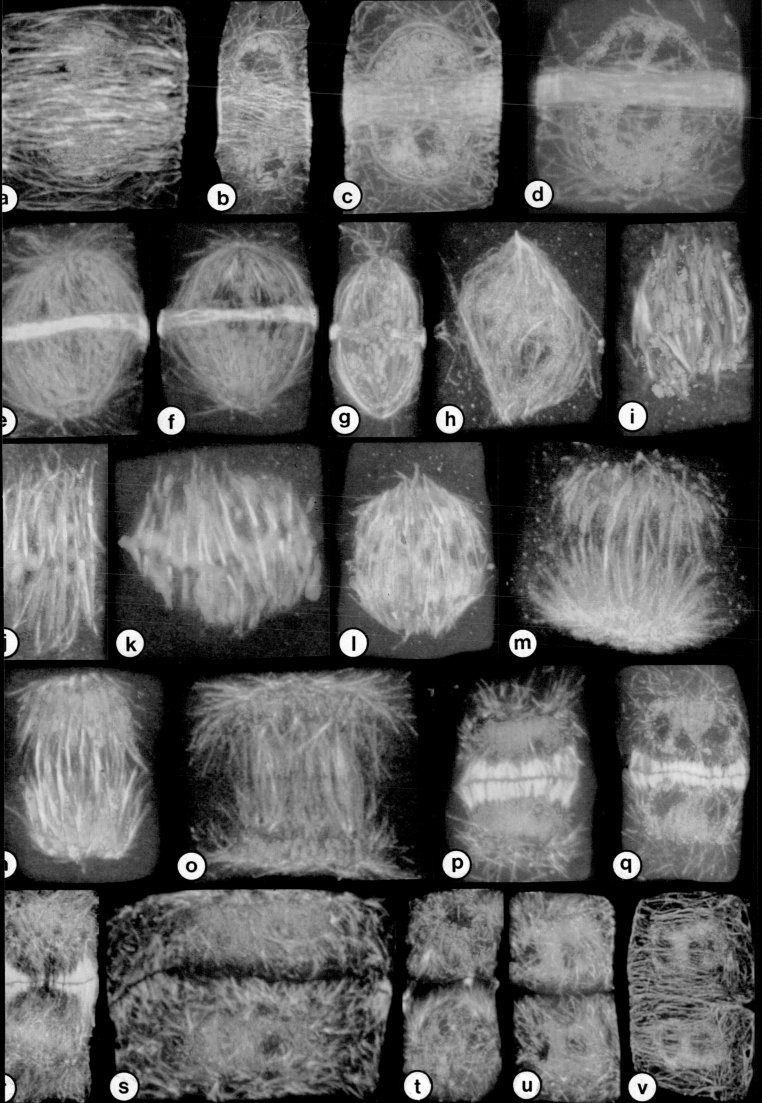

38   Mitosis in *Haemanthus* Endosperm Cells

Introduction: In this set of micrographs, differential interference contrast microscopy has been used to obtain time-lapse illustrations of the course of mitosis in a single living endosperm cell from the blood lily, *Haemanthus katherinae*. Endosperm is a nutritive tissue derived from fusion of two central *primary endosperm cells* and one *sperm cell* in the embryo sac (see Plate 59b). After fertilisation, endosperm cell proliferation begins without concomitant formation of cell walls, generating *liquid endosperm* tissue around the developing embryo (see Plate 59d,e). Samples of endosperm fluid containing these wall-less cells can be removed from young fruits and spread on a microscope slide for sequential observations of mitosis. They have been used in classic studies of spindle and phragmoplast dynamics, but are in a small minority in plants in lacking a wall and in not making a new wall after mitosis. Plate 39, showing division in a walled cell, is therefore included for comparison; the material is less favourable optically, but it does include typical cytokinesis.

The cumulative total time from the first micrograph in the sequence is recorded at the end of each caption below. All the micrographs are at the same magnification (x700). Identification of some cell components is based on electron microscope studies of the same material.

Plate 38a-c *Prophase.* The sequence begins when the cell was at an advanced stage of prophase. The nuclear envelope (NE) forms a distinct boundary (see also nucleus N-1 in Plate 1, which is at a comparable stage of mitosis) between the nuclear contents (condensing chromosomes, CH; disintegrating nucleolus, NL) and a clear zone that had already developed around the nucleus. The clear zone is occupied by the microtubules of the prophase spindle, which is initiated outside the nucleus (see Plate 36b and Plate 37e-g). In this case the clear zone is 3-polar (asterisks in (b)), perhaps because the cell has been flattened slightly for observation. The small protrusions of the nuclear envelope at the poles are where microtubules converge from different directions. As prophase progresses, the 3-polar condition gives way to the normal bi-polar division figure (c). The cytoplasm contains numerous small vacuoles (V), but other components are not readily identifiable except for a large mitochondrion (M) which can be seen in (a). Times: (a) 0; (b) 15 min; (c) 22 min.

Plate 38d-f *Prometaphase-metaphase.* Following rupture of the nuclear envelope (between (c) and (d)), a normal bi-polar spindle forms (poles at asterisks in (d)). The spindle fibres are as yet barely discernible at this level of resolution. Oscillations of chromosomes in the pole-to-pole axis during prometaphase (d) and (e) result in the orientation of their kinetochores (points of attachment to spindle fibres) at a balance point at the equator (E-E), thus establishing the metaphase stage (f) (see also Plate 37i-k). Kinetochore fibres (arrowheads) are particularly prominent in the upper half spindle. The chromosomes have been double since S-phase, consisting of chromatids twisted around one another (a). The sister chromatids now gradually unravel (d) and (e), prior to their separation in the next stage of mitosis. Times: (d) 1h 2 min; (e) 1h 18 min; (f) 1h 40 min.

Plate 38g-i *Anaphase.* Movement of the sister kinetochores to opposite poles gives the trailing arm chromosome configurations typical of early- (g) and mid- (h) anaphase. At these stages the kinetochore fibres (arrowheads) are still visible (see also Plate 37l,m,n). By late anaphase (i) the chromosomes have begun to coalesce at the poles. Note the rapidity of the movements of chromosomes during anaphase, as compared with the slower movements during prometaphase and metaphase. Phragmoplast fibres now develop at the centre of the spindle, and this activity results in lateral splaying movements of the trailing chromosome arms. Compare (i) and (h) to see how they fan out from their previous alignment along the axis of the spindle. Golgi stacks (G) begin to invade the phragmoplast region. Times: (g) 1h 50 min; (h) 1h 56 min; (i) 2h 7 min.

Plate 38j-l *Telophase - early cytokinesis.* Progressive condensation of the two chromosome masses results in the formation of two daughter nuclei. Nuclear envelopes are restored (NE in (l)). A comparable stage is shown in nucleus N-2 of Plate 1, where a newly-regenerating nucleolus lies amongst the uncoiling chromosomes. The nucleolus is not shown here, though it is present at a different plane of focus. Phragmoplast fibres are more clearly resolved between the two nuclei (F in (j)) and particles begin to accumulate along the equator, giving rise to the cell plate (CP in (k) and (l)). Golgi bodies (G) in side view and face view ((k) and (l) respectively) are prominent among the fibres. A very long and attenuated mitochondrion (M) lies across the cell plate in (l). Immunofluorescence microscopy of microtubule distributions in these cells has shown typical phragmoplast arrays, despite the fact that the raft of cell plate vesicles merely makes plasma membrane partitions between the progeny, and not a solid wall (compare Plate 39). There are relatively few situations in higher plants where mitosis is not followed by typical wall deposition - in meiosis, in embryo sacs, the division that forms sperm cells in pollen, and divisions in liquid endosperm. In these cases phragmoplasts consistently follow mitosis and cytokinesis is completed by plasma membrane formation. However, they often show little or no wall deposition. It is probably significant that no preprophase band of microtubules develops in these situations, presumably because their developmental program does not include insertion of a new cell wall at a predetermined division site in the parental cell. Times: (j) 2h 13 min; (k) 2h 22 min; (l) 2h 40 min. All micrographs kindly provided by A.S. Bajer.

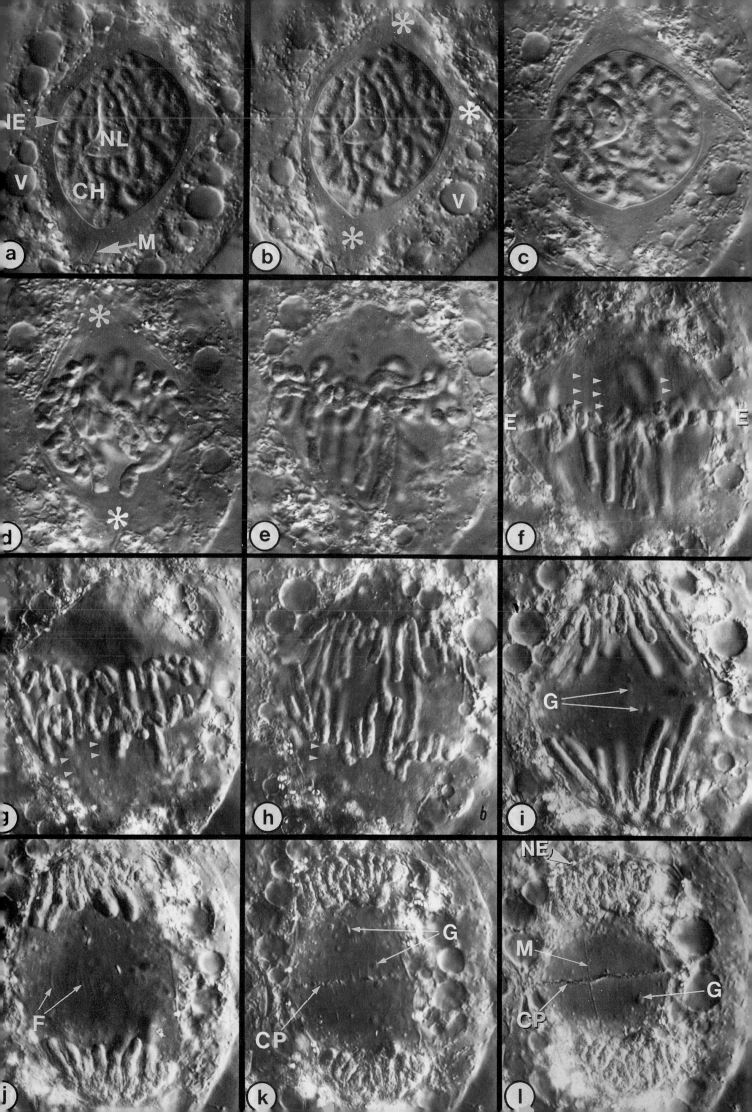

39   Mitosis in *Tradescantia* Stamen Hair Cells

Introduction: The advantages of using stamen hair cells of the plant *Tradescantia* for observing division in plant cells were exploited by Strasburger over a hundred years ago, and no better material has been found since. Stamens are dissected from 3mm long buds and young hairs are mounted in a simple culture medium. Each hair is a simple filament of transversely dividing cylindrical cells. Unlike *Haemanthus* endosperm cells (Plate 38), these are typical walled cells. They prepare a division site before mitosis and initiate a new wall as a cell plate within a phragmoplast. After centrifugal growth, the new wall fuses with the parental walls at the predetermined division site. The first three rows in this plate show time lapse differential interference contrast micrographs of mitosis in a single cell (elapsed times in minutes from the first image are given; all x1,300). The bottom row (b-k) shows confocal images of microtubules and actin in living stamen hair cells that were microinjected with fluorescent tubulin or fluorescent rhodamine phalloidin.

Plate 39a Prophase: The sequence starts just after the nuclear envelope broke down at the end of prophase ("time 0", equivalent to Plates 37g, 38a-c). The chromosomes are condensed and clear zones are seen at the poles, where the spindle is forming outside the nucleus on an axis at right angles to the plane of division.

Prometaphase-Metaphase: Prometaphase movements commence as soon as the nuclear envelope breaks (3,5), and continue until a metaphase plate has been established (20). The chromosomes capture kinetochore microtubules and are pushed/pulled into hairpin shapes as they are moved towards their equilibrium positions on the metaphase plate. In this cell the metaphase plate is oblique (between arrows at 20 min). The orientation of the metaphase plate can vary widely from the plane of division. Some cells routinely have very oblique spindles and undergo reorientation during anaphase.

Anaphase: Anaphase is well under way 28 minutes after the first picture. The chromosomes, seen to be double at the 5 and 20 min stages (e.g. small arrows) have now split into their constituent chromatids, which are moving to their respective spindle poles. Anaphase is rapid. Four minutes later (32) the chromatids have separated into two groups (equivalent to Plates 37m and 38i, except that in the latter the chromosome arms splay out, a feature not seen in walled cells). The kinetochore fibres continue to shorten and pull the kinetochore regions into a tight polar bunch (compare the rounded polar profiles at 40 and 42 minutes with the planar group of kinetochores that led the way towards the poles at 32 minutes).

Telophase and Cytokinesis: Aggregation of vesicles in the early phragmoplast becomes visible between 40 and 42 minutes (equivalent to Plate 37o). Further coalescence and outward growth of the new cell plate takes about 25 minutes (equivalent to Plate 37p,q,r). When the new cross wall completes its fusion with the parental walls, the formerly wrinkled cell plate (see 55,60,64) suddenly flattens (69). This does not happen unless the plate fuses in the correct place (at the division site), suggesting that factors donated from the site aid maturation of the new wall. The position of the division site has been visible since anaphase, marked by a thin raft of cytoplasm across the centre of the cell (see 32). This raft is a simple example of a *phragmosome*. Phragmosomes are more conspicuous in larger and more vacuolated cells.

Centrifugal outgrowth of the phragmoplast in this symmetrical transverse division is generally towards the division site. Marked curvatures or directed growth of the cell plate are not required (cf. Plate 40). Nevertheless there are oscillations (e.g. at 45 minutes the cell plate is noticeably tilted) which are continually corrected. Spatial guidance of cell plate growth is much more obvious if the mitotic figure is displaced experimentally, e.g. by centrifugation. Then it can be seen that the division site is the target of a guidance mechanism that directs the edges of the extending phragmoplast very precisely to the site that was predetermined before mitosis, when the preprophase band of microtubules was laid down.

Plate 39 b-e This stamen hair cell was microinjected with fluorescent-labelled tubulin protein, which became incorporated into microtubules in the course of a few minutes. (b,c) are surface and mid plane views very early in preprophase band development; (d,e) are similar views 11 minutes later, after a distinct preprophase band has formed. These images were the first direct views of preprophase band formation in a living cell. x1,100.

Plate 39 f,g These two pictures are equivalent to (b,d) but show F-actin microfilaments (after microinjection of rhodamine labelled phalloidin) in a band at the division site in a living cell. They were taken 5 minutes apart and show build up of actin at the division site in preprophase, at the same time and in the same orientation as the microtubules of the preprophase band. x1,100.

Plate 39 h-k When the cell enters prometaphase, the microtubule and actin bands at the division site both break down. Microtubules thereafter vacate the cell cortex completely (Plate 37), but actin microfilaments persist except at the division site ((h) - prophase; (i) - prometaphase), which thus becomes distinctively marked as a zone of actin depletion, relative to the rest of the cell surface. The actin depletion zone remains throughout mitosis, and by the time of cytokinesis it is still there, perhaps serving as a distinctive target for outward growth of the cell plate to the division site. (j) and (k) show surface and mid planes of focus of a cell roughly equivalent to the 52 minute stage in (a). Arrows mark the actin depletion zone; actin is also faintly visible in the phragmoplast. x1,100.

(a) kindly provided by P. Hepler; (b-k) kindly provided by A. Cleary; (b-e) and (h-k) reproduced by permission from *J. Cell Sci.* **103**, 977-988, 1992.

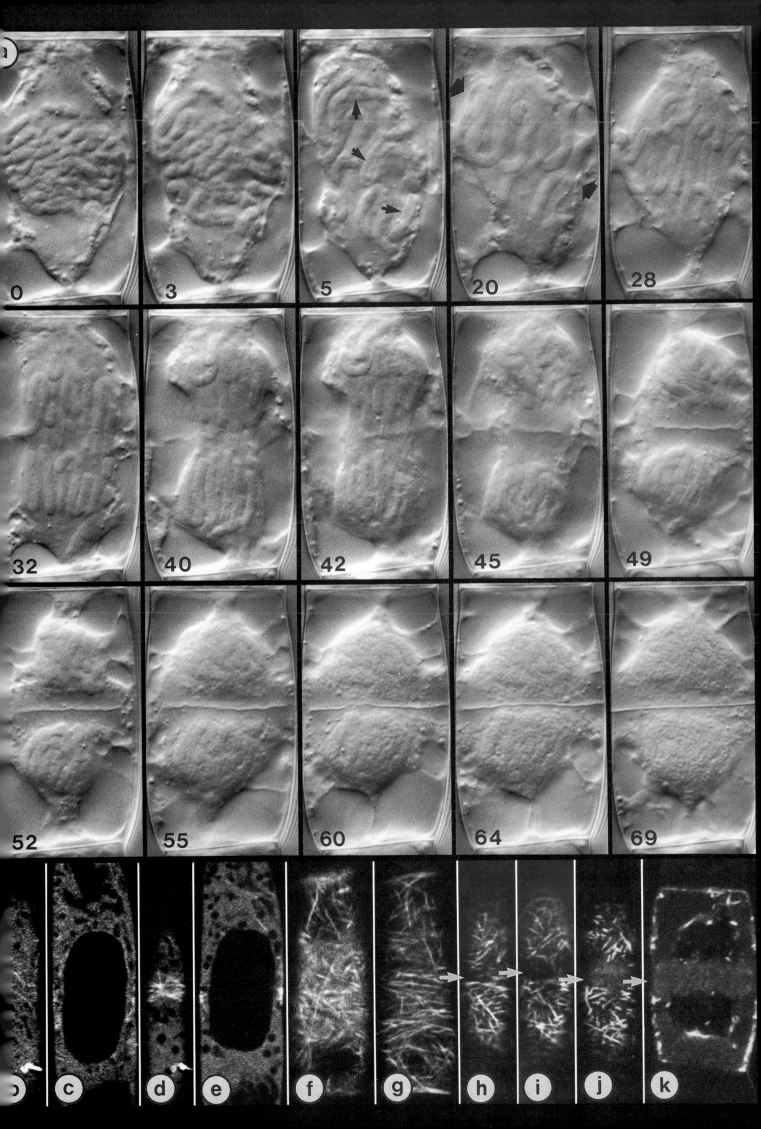

40 The Preprophase Band in Asymmetrical Cell Division (Subsidiary Cell Formation in Stomatal Complexes)

Introduction: Plate 37 illustrates development of the preprophase band (PPB) of microtubules (MTs) in root tip cells that divide transversely and symmetrically. The present plate illustrates an example of a very asymmetric cell division, part of a precisely programmed sequence that forms *stomatal complexes* in grass leaves. First, certain cells in the leaf epidermis divide transversely (but asymmetrically) to produce the *guard mother cells*, i.e. the cells that will give rise to the two *guard cells* of each stoma. Each guard mother cell exerts a strong polarising influence over the epidermal cells that flank it to right and left along the leaf axis. They become *subsidiary mother cells*. Their nuclei become attracted towards the guard mother cell and they are induced to divide beside it. A small lens-shaped cell is cut off on either side of the guard mother cell, to become the *subsidiary cells* of the stomatal complex. Then the guard mother cell divides symmetrically in the longitudinal axis of the leaf to form the two guard cells, between which the stomatal pore later develops. The division that produces the subsidiary cell is shown in this Plate. All of the micrographs are from sugar cane leaves.

Plate 40a   This survey electron micrograph shows a guard mother cell in the centre with "before" and "after" stages of subsidiary cell division above and below it. The long axis of the leaf is from left to right (as also in (d)-(g)). At the bottom is a recently-formed subsidiary cell. Its cell wall attaches to the walls of the cell file in which the guard mother cell lies, just outside the boundaries of the guard mother cell itself (arrows). The position of this site of attachment is very consistent from one stomatal complex to another and indeed from one plant to another. The vacuole (v) at the lower edge of the picture lies in the very large sister cell of the subsidiary cell. At the top is the nucleus of a subsidiary mother cell. It has been attracted to lie alongside the guard mother cell but has not yet undergone mitosis. Nevertheless the future site of division has already been pinpointed. At the site where the subsidiary cell wall will fuse with the parental wall, there are bundles of PPB MTs, seen in cross section (brackets). Comparison of this "before division" stage with the "after division" stage at the bottom shows how accurately the position of the PPB predicts the future site of division. Note that the two bundles of PPB MTs seen here are just two profiles of a continuous three dimensional band around the cell surface, specifying where the new cell wall will be inserted. x11,000.

Plate 40b,c   These two pictures show the two profiles of a PPB in a cell at a similar developmental stage to the one at the top of (a). Some PPB MTs are closely associated with the plasma membrane. Those lying deeper in the cytoplasm show quite precise hexagonal packing arrangements in places. The stage of development of this PPB corresponds to the "mature" PPBs illustrated in Plate

37d,e, where the MTs are too close to one another for immunofluorescence microscopy to distinguish the individuals. The PPB disappears during mitosis (see Plate 37) but leaves a "memory" to guide the growth of the new cell plate at cytokinesis (see below). Both x46,600.

Plate 40d   This micrograph shows metaphase of the subsidiary cell division, with the chromosomes in a distinct metaphase plate. In this and the remaining pictures a portion of the guard mother cell appears at the lower edge. The walls at left and right of the guard mother cell show where the subsidiary cell wall will be inserted (open arrows). As in root tip cells at metaphase (Plate 37k), no PPB MTs remain, though some MTs of the mitotic spindle are visible (arrows). x10,300.

Plate 40e   *Anaphase.* Spindle MTs are visible (arrows) but no MTs are to be seen at the former PPB site (open arrows). x11,900.

Plate 40f   *Early cell plate development.* The nucleus that will become the subsidiary cell nucleus is closely appressed to the guard mother cell; a small part of its sister nucleus is just visible at the top of the micrograph. As in (d) and (e) the site of division (open arrows) has no PPB MTs. The new cell plate has been initiated as a raft of vesicles between the daughter nuclei. Initially it grows outwards from the centre of the spindle in a flat plane. Phragmoplast MTs are present (arrows). x10,000.

Plate 40g   *Late cytokinesis.* The new cell plate has now been guided round to attach to the parental wall at the site where the PPB formerly lay (arrows). The sequence (f)-(g)-(lower cell in (a)) shows progressive stages of cytokinesis, including the origin of plasmodesmata where strands of endoplasmic reticulum traverse the new wall (also Plates 36m,n, 44) and progressive decondensation of the nuclear chromatin after mitosis. The new wall is still immature in the lower cell in (a): it has not yet been consolidated by wall deposition and is still distorted. Later it will smooth out into its final shape (see also Plate 39a, 69 min stage). x13,000.

It seems that the division site, which is determined before mitosis and distinguished by the PPB, may have at least two roles. The first is to establish a mechanism that later will guide the growth of the phragmoplast to the division site. This is especially clear in asymmetric divisions like that shown here. The second may be to direct local deposition of (as yet hypothetical) "wall maturation factors" into the parental walls at the PPB site, to remain there until they gain access to the new wall after it has been guided to and fuses with the parental walls at the predetermined site. They then help to consolidate the new wall. One piece of evidence for this second role comes from experiments in which a cell plate is made to fuse at a "wrong" site, e.g. by centrifugal displacement. In such cases it usually fails to mature properly.

Micrographs kindly provided by C.H. Busby.

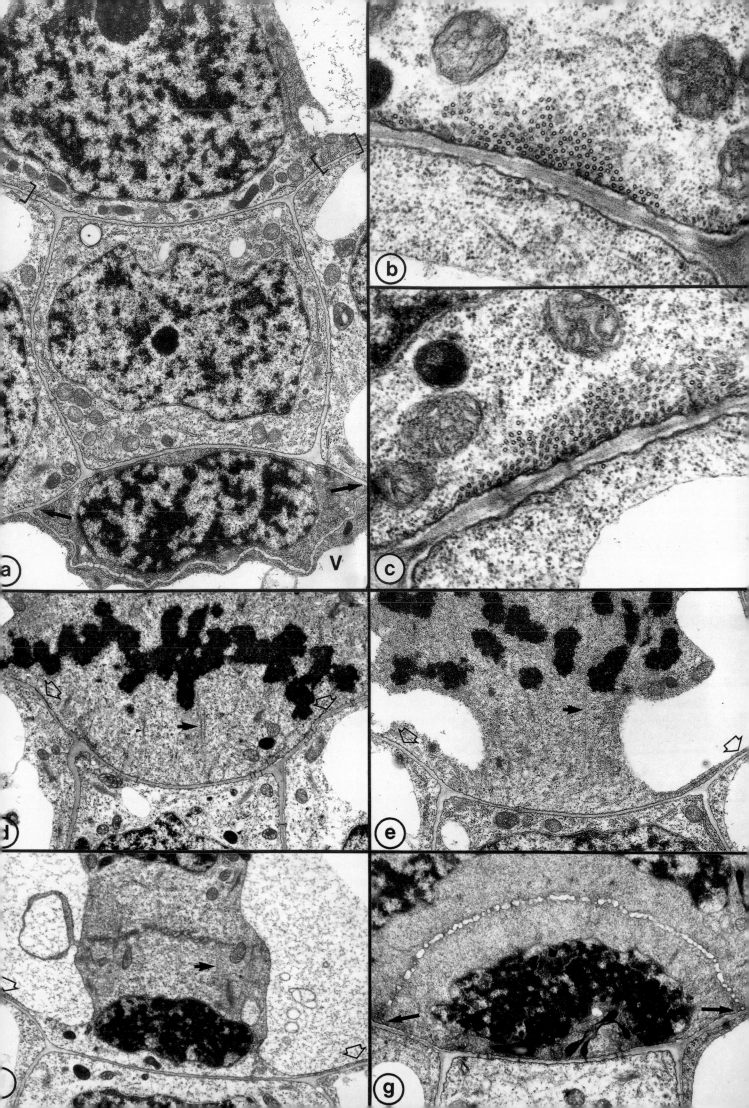

41 Mitosis: Prophase

Introduction: A typical round of cell division generates two new cells from one parent. Each of the progeny receives a complete set of genetic information, partitioned with precision following exact duplication in the parent, and a set of cellular components, partitioned more simply by sharing out the parental cytoplasm. Although they are genetically identical, the fate of the progeny may or may not differ. Both progeny may enter new cycles of cell division, as happens in parts of root and shoot meristems. One of the progeny may remain meristematic while the other leaves the cell cycle to embark on a program of differentiation, as happens at the margins of meristems, or in the vascular *cambium*, which is a meristem consisting of a single layer of cells that generates xylem and phloem to its inside and outside. Alternatively, both progeny may differentiate, as for example in *meristemoids,* restricted zones of dividing cells that arise in non-dividing tissues and generate restricted multicellular systems such as stomatal complexes.

Plates 37-40 have illustrated some of the many partial processes that are integrated spatially and temporally during cell division. These events are initiated and the orderly progression coordinated by particular gene products that act in sequence to drive the succession of mitotic events. In nearly all cases the parental cell must grow by a threshold amount before it can embark on division. At this stage one of the most fundamental genes in all organisms, the "start" gene for cell division, is activated. It triggers the S-phase of the cell cycle (see Plate 34), during which DNA is replicated. Further events then occur in the cytoplasm, during the G2 phase, in preparation for mitosis. One of these, in all organisms, is a check to ensure that DNA replication has been completed. In plants another event is the establishment of the division site, marked by the formation of the preprophase band, consisting of cortical actin and microtubules. A complex of two proteins, one a mitotic cyclin and the other a protein kinase, is then brought to peak activity, triggering prophase, the first phase of mitosis itself. Prophase is characterised by condensation of chromosomes (by now double) and a major change in cytoskeletal organisation, namely production of microtubules around the nucleus. Although they are still outside the nucleus, these microtubules are the precursors of the mitotic spindle, and become aligned in the future pole-pole axis.

Plate 41 displays electron micrographs of ultra-thin sections of cells during and at the end of prophase, complementary to the light micrographs of Plates 36b,c, 37d-h, 38a-d, and 39a(0,3). Plates 42-44 continue the sequence of illustrations of mitotic phases. Here and in Plates 42-44, unless stated otherwise, all of the sections are longitudinal with respect to the cell axis. Where possible, the micrographs are oriented so that the spindle poles lie at the top and bottom of each figure.

Plate 41a The major feature of this prophase nucleus is its chromatin (CH), condensed into the form of chromosomes. The nucleoplasm, still enclosed by the intact nuclear envelope (NE), shows little structure at this magnification. The nucleolus (NL) undergoes complete dispersal in most mitoses, starting before the nuclear envelope breaks at the end of prophase. An initial stage of dispersal is seen here, with the granular and fibrillar zones accentuated as nucleolar transcriptional activity (see Plate 6) diminishes. Nucleoli are restored later in mitosis (Plate 43). There is no obvious clear zone around the nucleus, although many pre-spindle microtubules (not seen at this low magnification) are forming around it. Proplastids (P), mitochondria (M) and vacuoles (V) are all lying close to the nuclear envelope. x6,500.

Plate 41b The area enclosed by the rectangle in (a) is shown at higher magnification in (b). NL marks a mass of nucleolar fibrils, and nucleolar granules are at upper left. Part of the chromosome (CH) is also included. The background nucleoplasm contains no microtubules at this stage (cf. below). x28,000.

Plate 41c-e These micrographs illustrate the breakdown of the nuclear envelope, the event that defines the end of prophase. It is triggered by the activated protein kinase referred to above. Areas enclosed by the rectangles in (c) are shown enlarged in (d) and (e). Breaks in the nuclear envelope (NE) (see asterisks in (c) and (e)) have released fragments of endoplasmic reticulum-like cisternae, their origin apparent from the retention of nuclear envelope pores in them (NP in (e)). Although the nuclear envelope has only been partially removed, the nucleoplasm between the chromosomes (CH) has already been invaded by cytoplasmic ribosomes, mostly in the form of polyribosomes (compare (c) and (d) with (a) and (b) above). Vesicles (VE) also appear in the former nucleoplasm (note the presence of Golgi stacks (G in (c)) outside the disintegrating envelope). Components of the future spindle have made their first appearance in the nucleoplasm. Microtubules (MT in (d) and (e)) are now evident near breaks in the nuclear envelope (e) as well as in the outer region of the nucleoplasm (e) and in central regions (d). The nucleolus (NL in (c) and (d)) is more dispersed than in (a).

One of the most striking changes is in the nature of the nuclear envelope. When intact, it bears pores, and ribosomes are bound only to the cytoplasmic face of the outer membrane (Plates 5d). It has been caught here in an intermediate state in which it more closely resembles normal cisternae of rough endoplasmic reticulum. Thus ribosomes are now seen on both the inner (solid arrows in (e)) and the outer membrane. Pores are still present. A similar state recurs at telophase of mitosis, and will be discussed in relation to the cycle of breakdown and resynthesis in Plate 43d. (c) x16,000; (d) and (e) x50,000. (a) - (c) all from root tip cells of *Vicia faba.*

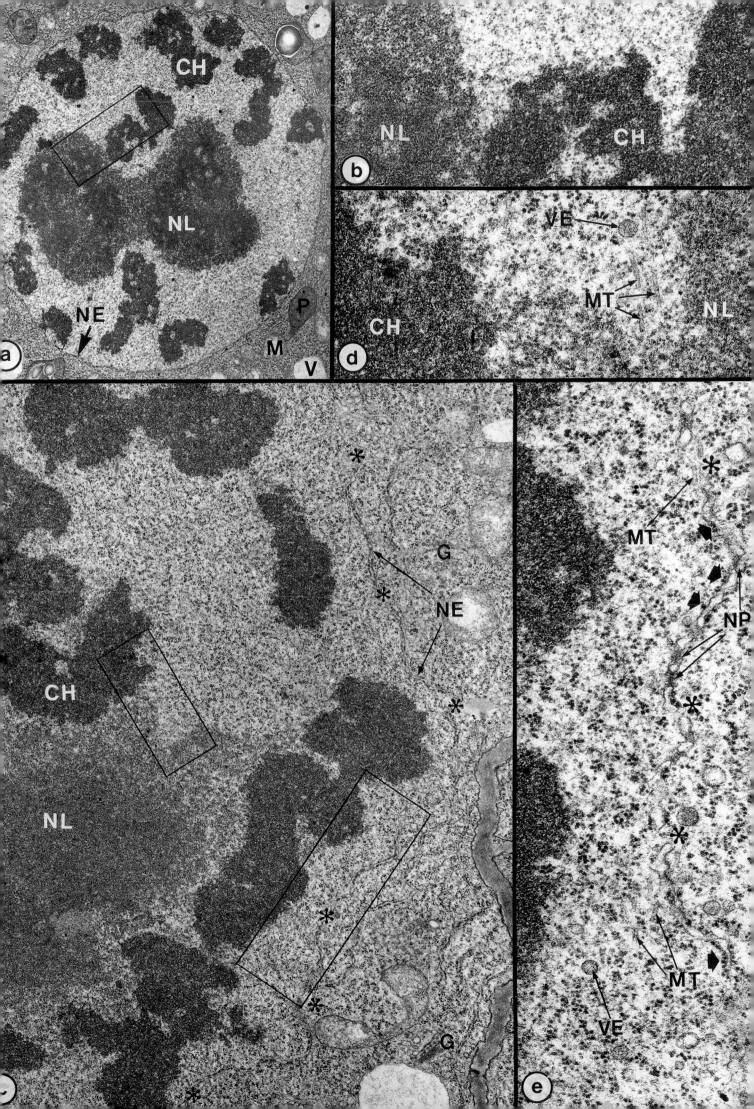

42 Mitosis: Prometaphase and Metaphase

Introduction: Whereas the transition from the G2 phase of the cell cycle to prophase of mitosis is gradual, that from prophase to the next stage, *prometaphase*, begins abruptly. The nuclear envelope ruptures suddenly, probably by phosphorylation-induced disassembly of a nuclear cytoskeletal layer that normally supports it. Major changes in the spatial arrangement of the chromosomes commence, driven by microtubules which now pass among the chromosomes from their earlier position in the extra-nuclear prophase spindle. Each chromosome was duplicated before mitosis in the S-phase of the cell cycle, and now consists of two *chromatids* joined together at specialised tracts of repeated DNA sequences, the *centromere* regions. Two kinds of protein interact with centromeric DNA: (i) proteins that hold the sister chromatids together until the metaphase-anaphase transition, and (ii) proteins that attach chromatids to microtubules and affect the stability of the microtubules. The latter proteins constitute the *kinetochore*, the site of attachment of chromatids to the mitotic spindle.

Spindle microtubules are organised in two interdigitating populations, one emanating from each pole. Animal cells generally have discrete spindle pole structures, but plant cells do not. Despite this the microtubules emanating from each diffuse polar zone are arranged with opposite molecular polarity (see Plate 31). Kinetochore proteins have the capacity to bind the "fast-growing" end of microtubules, i.e. the end furthest from the pole. Kinetochores vary in size - the minimal size allows for "capture" of just one microtubule per chromatid; the larger examples may capture numerous microtubules. Kinetochores may also initiate polymerisation of additional microtubules. In living cells, if just one kinetochore in a pair of chromatids captures microtubules, the chromatid pair moves rapidly towards the pole from which those microtubules have come. Then, when the sister kinetochore captures its microtubules from the opposite spindle pole, the chromatid pair undergoes a series of oscillations between the two poles. Eventually a balance point is reached with the pair of kinetochores lying midway between the two poles. Chromosomes will not associate stably with the spindle unless both chromatids are attached *via* their kinetochores to opposite spindle poles.

By the time all chromatid pairs have undergone their prometaphase adjustments, they are aligned, more or less side-by-side, in the *metaphase plate*. Metaphase is a relatively prolonged stage of mitosis - indeed mitosis does not usually proceed until all chromatid pairs have become properly aligned. Meanwhile they oscillate in a state of tension, as can be shown by severing a kinetochore bundle - whereupon the surviving bundle pulls both chromatids towards its pole. The state of "equilibrium-under-tension" establishes the starting conditions for chromatid separation during anaphase (Plate 43).

Plate 42a This micrograph continues the series showing mitosis in *Vicia faba* root tip cells, started in Plate 41a and c. It shows part of a group of chromosomes, liberated from a nucleus and in prometaphase. Microtubules are now present in extensive arrays (MT) among the chromosomes (CH), which have not yet become aligned on the equator of the division figure. The nucleolus (NL) is now so dispersed as to be scarcely detectable. (stage corresponding to Plate 36d,e, 37h, 38d, 39a(5)). x11,000.

Plate 42b, c and d illustrate stages of mitosis in cells in root tips of white lupin (*Lupinus albus*), which has smaller cells and chromosomes than the broad bean (*V. faba*) used in Plates 41 and 42a.

Plate 42b As was the case in (a), the chromosomes shown here were fixed during prometaphase movements, and hence lie at different, non-equatorial, levels of the division figure. The microtubule (MT) system has developed considerably and now consists of bundles oriented in the pole-to-pole axis. Those seen in this micrograph mostly disappear into masses of chromatin in chromatids (CH), i.e. they are kinetochore microtubules that have been captured by (and perhaps also initiated at) kinetochore regions (see also Plate 37i). x36,000.

Plate 42c This section at the edge of the equatorial region of the mitotic figure in a metaphase cell shows aligned, paired chromatids (CH) at the end of the prometaphase movements (equivalent to Plate 37k, 38e, 39a(20)). Kinetochore fibres (KF, hardly visible at this magnification) run from the kinetochores, oppositely oriented on the chromatid pairs, and mix with other, non-kinetochore, bundles of microtubules in the spindle, all passing towards the poles. Proplastids (P), mitochondria (M), Golgi bodies (G), lipid droplets (L) and vacuoles (V) are in general excluded from the spindle region and lie between it and the cell wall. x8,000.

Plate 42d Kinetochores vary greatly in structure. In these simple types there is just a matrix (K) in which the microtubules (KM) terminate. Other microtubules (MT) penetrate between the chromosomes (CH), and probably are examples of pole-to-pole microtubules (as distinct from pole-to-kinetochore). The insets present the alternative view of a kinetochore, seen in sections cut in the plane at right angles to that of the main micrograph, and at a position equivalent to the level of the arrowheads from the label K. The material was a dividing *Chlorella* (unicellular alga) cell. The upper inset shows transversely sectioned microtubules (ringed), one single, and one pair, in each case surrounded by fuzzy material. In the adjacent section (lower inset) of precisely the same area, the microtubules are no longer visible, and more of the fuzzy kinetochore or chromosomal material is included. The microtubules clearly are of the kinetochore type, and the two sections must have spanned the microtubule termini in or on the chromatids. x36,000; insets x45,000 (insets kindly provided by A.W. Atkinson, Jr.)

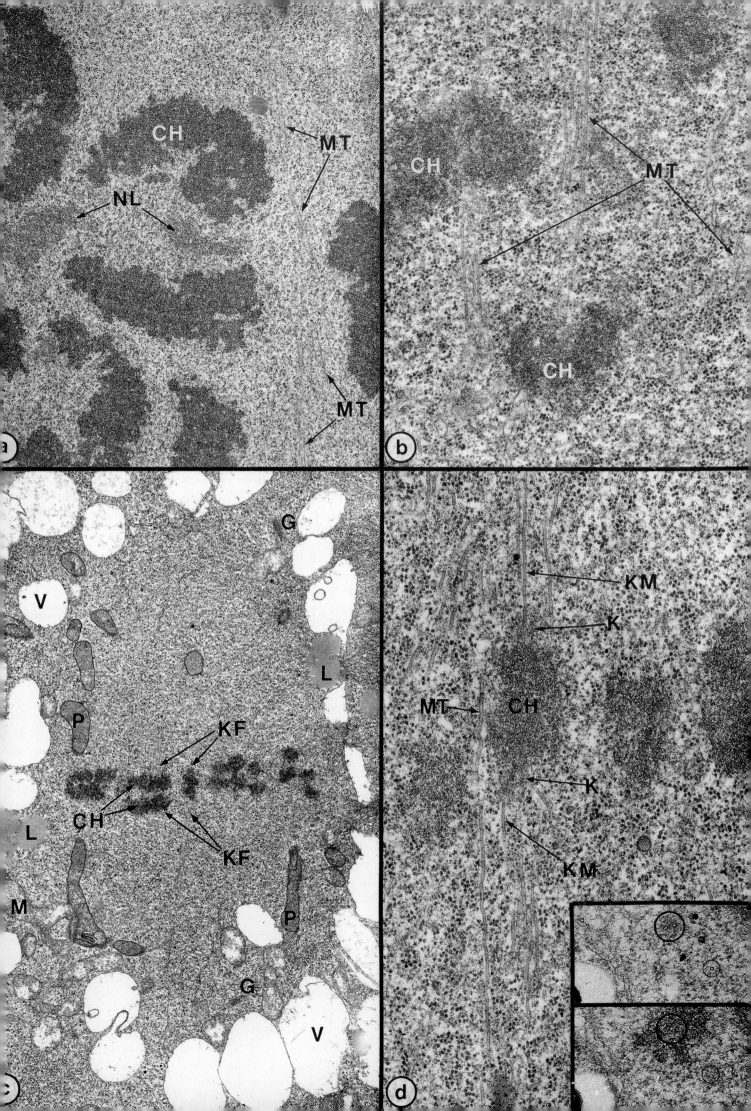

43   Mitosis: Anaphase - Early Telophase

Introduction: By the end of metaphase, the chromatid pairs are aligned on the equator of the division figure, still not separated, but connected to the opposite poles by tensioned microtubule bundles. The next step is abrupt, like the earlier, sudden rupture of the nuclear envelope. The linkages that hold the sister chromatids together are severed simultaneously in all chromatid pairs. Separation does not involve microtubules, but it lets the microtubule-driven poleward movements of anaphase begin.

Anaphase is a rapid process. In Plate 39a, *Tradescantia* chromatids are seen to move to the poles at about 1μm per minute. In most plants the greatest contribution to separation of sister chromatids comes from shortening of the kinetochore bundles (see Plate 37). This is known as *anaphase-A*. It is thought that tubulin subunits are removed from the microtubule attachment regions at the kinetochores, and that kinetochores can also be moved along the kinetochore bundles by microtubule (or other) motor proteins. The kinetochores therefore approach the mitotic poles, their chromatid arms trailing behind. Indeed the kinetochores often congregate into a confined region in the middle of the diffuse pole (Plate 37n, 338i,j, 39a(40,42)), indicating that the pole possesses some form of discrete identity, despite its lack of an equivalent of the centrioles that define the centre of the pole in most animal cells. Another contribution to separation comes from a general separation of the poles, i.e. a lengthening of the spindle, generated by inter-microtubule sliding in pole-to-pole bundles. This is known as *anaphase-B*. Its contribution varies considerably, and scarcely exists in many plant cells where the spindle occupies most of the available space within the walls.

The micrographs in this plate show selected ultrastructural details of anaphase-telophase stages, particularly of restoration of nucleoli and the nuclear envelope, and should be viewed in conjunction with images of live cells and microtubule immunofluorescence in Plates 36i-k, 37l-n, 38g-i, 39a(28-40).

Plate 43a This mid-anaphase cell from a white lupin root tip shows parts of several chromatid arms, caught within the section at various positions between the equator (which runs approximately left to right) and the spindle poles (at top and bottom). Two recently-separated sister chromatids, each carrying a nucleolar organizer region of chromatin (NO), lie close to each other at the left hand side of the spindle. The nucleolar organizers are clearly different from the chromatid material on either side of them. Although light microscope staining reactions indicate that the concentration of DNA is relatively low in them, they carry the hundreds of repeats of rRNA genes (Plate 6). These genes are inactive during mitosis. The whole spindle region, filled with numerous ribosomes and portions of sectioned spindle microtubules (MT), is surrounded by many cisternae of rough

endoplasmic reticulum (ER), outside which lie Golgi bodies (G), mitochondria (M), plastids (P), and vacuoles (V). x8,000.

Plate 43b, c Later in anaphase, when the chromosomes begin to coalesce at the polar regions (upper parts of b and c), each chromosome becomes surrounded by a zone of granular material (large arrows in b), separated from the chromatin by a narrow electron-transparent space. Micrograph (b) shows a trailing chromosome arm possessing a nucleolar organizer region (NO), and a further example of the same structure is included in (c). Spindle microtubules (MT) are seen in (b). Note also the vesicles (VE) amongst the trailing arms. Elements of endoplasmic reticulum modified by the development of nuclear pores (NP) are also present close to the chromosomes. *Vicia faba* root tip, (b) x16,000; (c) x10,000.

Plate 43d By early telophase a nuclear envelope with pores (NP) invests most of the coalescing chromosomes. The inset shows a portion of the new nuclear envelope in more detail, including a pore (large arrowhead). Where the reassembling envelope is not closely appressed to chromosome material, ribosomes and polyribosomes can be seen on both its inner and outer surfaces. Some of the ribosomes on the future inner membrane are arrowed. They will disappear from the inner face of the new envelope when it has been completed, but at this stage the fragments of nuclear envelope resembles those seen soon after the end of prophase (Plate 41e). It is not clear whether such fragments persist throughout metaphase and anaphase to contribute to the re-forming envelope at telophase, or whether the prophase fragments lose their pores and merge with the general endoplasmic reticulum. In animal cells there is evidence that fragments persist in a form that is rendered unable to assemble onto the chromosomes by protein phosphorylation. Dephosphorylation later allows them to bind to the telophase chromatin, where they fuse with each other and gradually reassemble a nuclear envelope. Additional pore complexes are inserted as the surface area of the nucleus increases during decondensation of the chromosomes.

The nucleolar organizer region (NO) has started to expand at its peripheral regions, differentiating an outer granular layer (G). Further expansion leads to regeneration of the nucleolus, as in nucleus N-2 in Plate 1, which is from the same material as the present micrograph (*Vicia faba* root tip), but is at a slightly later stage of telophase, with a larger nucleolus and with the chromosomal material beginning to de-condense towards the interphase condition. Growth of the nucleolus back to its active, interphase condition involves resumption of transcription by the rRNA genes and appearance of nucleolar materials on the surface of the chromosomes. These coalesce with the expanding nucleolar organiser region. x32,000, inset x52,000.

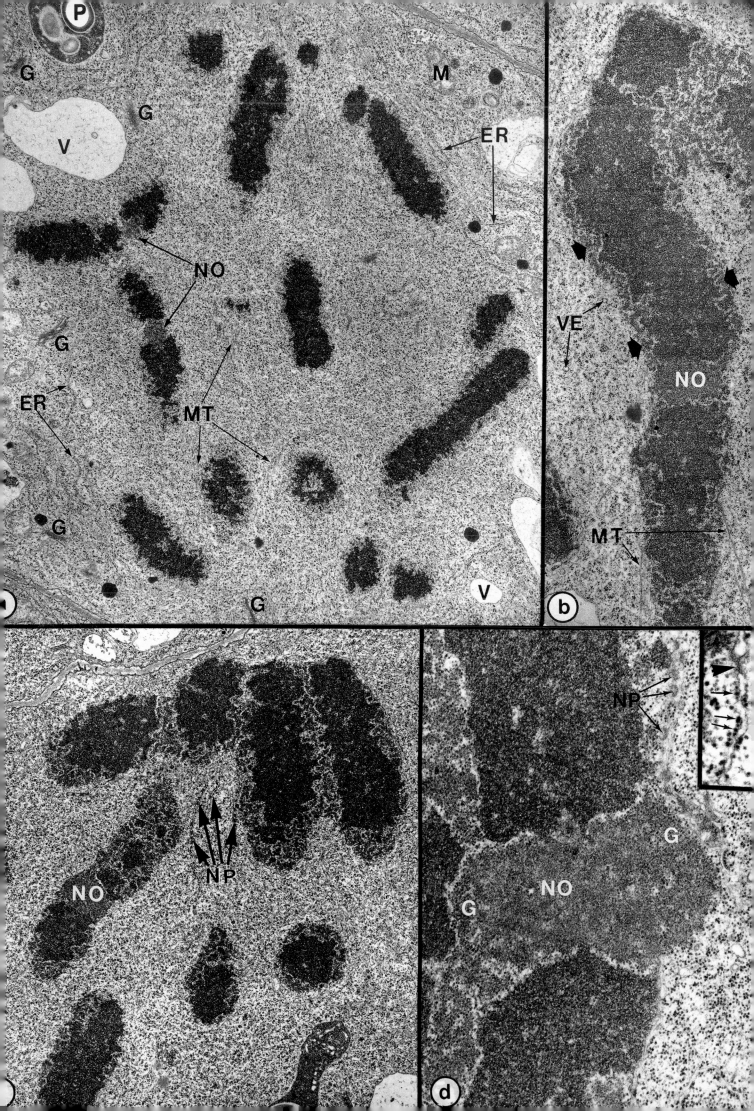

44   Mitosis: Telophase and Cytokinesis

Introduction: Mitosis, the partitioning of the genetic material, is basically similar in plants and animals. Cytokinesis, the partitioning of the cytoplasm, is very different. The differences arise from the different properties of the extracellular matrix in the two kingdoms. In animals it is generally pliable and cells can undergo reversible alterations in shape. Cytokinesis is accomplished by an acto-myosin based constriction, starting at the plasma membrane and working inwards. The progeny readjust their shape afterwards. In plants the mechanical constraint of the cell wall conditions the nature of cytokinesis. New walls have to be inserted into the fabric of pre-existing walls in particular sites and planes, with minimum deformation. This is achieved by initiating the new wall internally and then guiding its outward growth to predefined sites at the cell surface. Once the nuclear DNA has been precisely shared out, the emphasis therefore shifts back to coordination with preparations for cytokinesis that were made during the G2 phase, before control was taken over by the gene products that govern the sequence of events in the spindle apparatus.

The new internal wall is assembled in the phragmoplast, a complex apparatus of endoplasmic reticulum, microtubules, actin and Golgi-derived vesicular inputs into the growing cell plate. The first evidence of coordination with previous preparations is that the spindle, which has thus far been untethered in the centre of the cell, becomes aligned with respect to the division site. This adjustment is often necessary in large cells, where the spindle may have to be rotated as much as 90° to start guiding the growing edges of the cell plate towards the division site. Even in small cells, ′spindle lengthening (anaphase-B) may have pushed the spindle into a diagonal position, which now has to be readjusted.

Plate 44a This off-centre longitudinal section of a *Beta vulgaris* root tip cell is equivalent in stage to Plate 36l-m, 37o, 38k-l, 39a(45). It shows two daughter telophase nuclei (N), each bounded by a nuclear envelope (NE) that is by now complete. Between the two lie phragmoplast microtubules (MT) and the developing cell plate (CP). The discrete vesicles (VE) of the young cell plate have in places begun to coalesce to form larger units (large arrow). Continuation of coalescence gives rise to the plasma membranes of the cross wall separating the daughter cells. A Golgi stack (G) is present beside one of the nuclei. x13,500.

Plate 44b Phragmoplast microtubules and a young cell plate are seen here at the same stage as in (a). Microtubules (MT) meet and end (large arrows) at the raft of vesicles (VE) that forms the cell plate. Microtubules on opposite sides of the plate have opposite polarity, and motor proteins associated with them are believed to function in bringing vesicles (small arrows) into the growing plate. *Vicia faba* root tip cell, x23,000.

Plate 44c This large, highly vacuolated, tobacco BY-2 cell growing in tissue culture illustrates the internal origin of the cell plate and the high density of endoplasmic reticulum cisternae that contribute to its organisation. The living cell was vitally stained with rhodamine-123 (see also Plate 8). The cell plate membranes and some other cell components stain brightly; the endoplasmic reticulum is comparatively faint. The telophase nuclei are out of the plane of focus in this confocal micrograph. x1,100, inset x1,800.

Plate 44d This is a face view of a cell plate - that is, in an ultra-thin section cut at right angles to that of (a). Due to its undulating contour, the plate passes into and out of the plane of section, so that part of the micrograph includes the plate, and part the neighbouring cytoplasm. Some vesicles (large arrows) remain as discrete spheres between the advancing cell plate and the side wall (CW) but most of the image consists of vesicles that have coalesced into a branched array of membrane - the future plasma membrane at the new cross wall. Numerous profiles of cross-sectioned phragmoplast microtubules (arrowheads) are seen among the coalescing vesicles and tubules of the cell plate. *Avena sativa* anther, x47,000.

Plate 44e,f,g Small segments of cell plates are shown here in side profile (as in a) in order to demonstrate certain points of detail. Regions of the coalescing membrane surface (black arrows in e, and just to the left of ER in g) bear 15nm thick coats resembling those of coated vesicles (Plate 15). It is not possible to say from a static image such as this one whether coated vesicles gave rise to the coated areas of the cell plate by fusing with it, or whether coated vesicles are being formed at the cell plate, and have been fixed in the course of endocytosis. A second point, illustrated in (e) and (g), is that cisternae of endoplasmic reticulum (ER) pass through gaps in the growing cell plate. This is the source of the axial desmotubule structures of primary plasmodesmata (PD, see also Plate 45). The extent of the entrapment of endoplasmic reticulum is more evident from the lower magnification light micrographs of (c) and Plate 36m-o.

A point of contact between the side wall (CW) of the dividing parental cell and the extending cell plate is shown in (f). The dark objects in the parent cell wall are remains of plasmodesmata, which became occluded when the edge of the cell plate fused with the parental wall. As in (e) and (g), the coalesced vesicles of the cell plate contain fibrillar material, the first sign of the primary wall that will separate the two daughter cells. Nearby in the cytoplasm on both sides of the plate are numerous vesicles (VE), amongst the phragmoplast microtubules (MT). Hemispherical profiles on the membrane of the cell plate (open arrows in e and f) suggest that vesicles such as those lying free (VE) have fused with and delivered their contents to the plate. *Vicia faba* root tip cells, (d) x67,000, (e) x55,000, (f) x55,000.

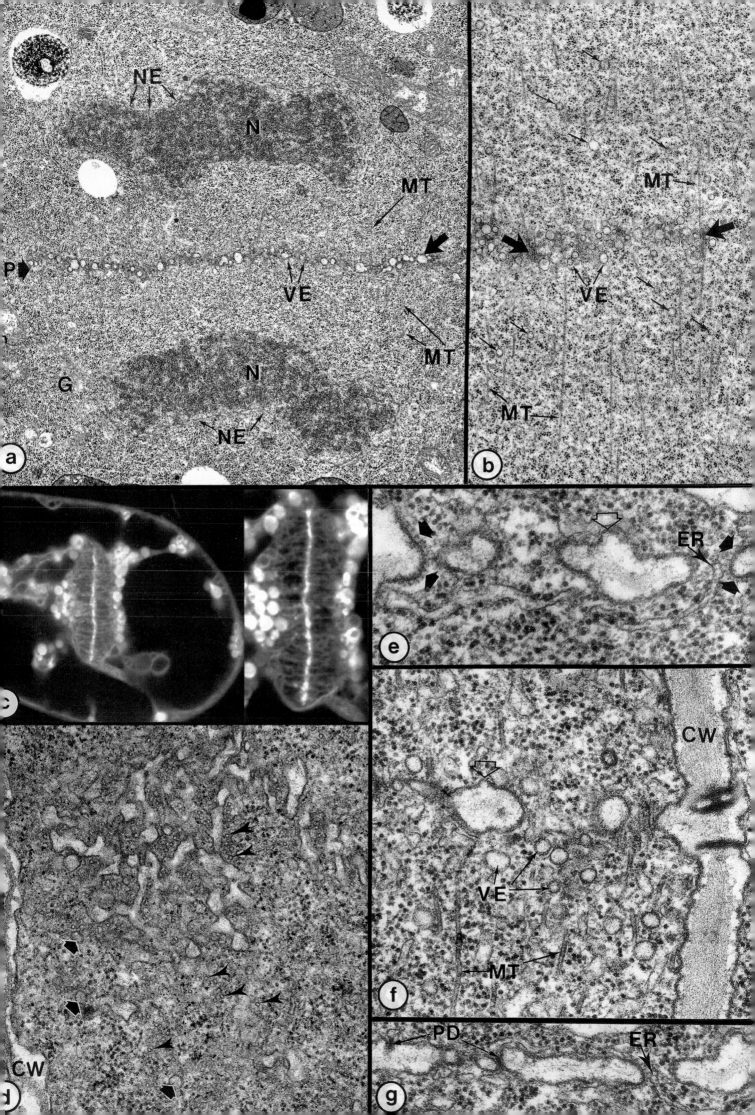

45 Plasmodesmata

Introduction: Plasmodesmata are tubular connections, 40-50nm in diameter, between adjacent plant cells. The plasma membrane is continuous from one cell to the next, forming the lining of the tube. An axial structure, the *desmotubule*, passes through each plasmodesma and is continuous with endoplasmic reticulum (ER) membranes at each end. Plasmodesmata convert plants from colonies of independent cells to domains of interconnected cells within continuous bounding plasma membranes. The domains are *symplasts*; the cells in them are *symplastically connected* and engage in *symplastic transport*. The extent of interconnection, and conversely the extent of isolation, depends on developmental patterns of formation and occlusion of plasmodesmata. Although plasmodesmata normally form in the cell plate at cytokinesis (*primary plasmodesmata*, see Plate 44), they can be sealed off later and they can also develop *de novo* in existing cell walls (*secondary plasmodesmata*). They can even arise between cells of different species in heterograft unions.

Plate 45a One plasmodesma is shown here connecting two cells in an *Azolla* root. Staining of the lipid head group layers of the bimolecular leaflet of the plasma membrane (arrows) and ER has been intensified by including tannic acid in the fixative. Continuity of the axial strand (desmotubule) with the ER is clear. The lumen of the desmotubule is wide enough to hold a few water molecules only (see also Plate 13d) and therefore is unlikely to be an effective channel for cell-to-cell transport. Its wall is probably ER membrane enriched by extra membrane proteins. The main pathway for intercellular transport is the annulus between the desmotubule and the cytoplasmic face of the plasma membrane. Most plasmodesmata support cell-to-cell movement of molecules up to a molecular weight of 800-900 Da (depending on solubility and shape). This low exclusion limit suggests that the cytoplasmic annulus is partly occluded, perhaps by proteins analogous to those in nuclear envelope pores, to micro-channels a few nanometres in diameter. x240,000.

Plate 45b End-on views of numerous plasmodesmata in a grazing section of a curved cell wall in an *Azolla* root meristem (x12,000). Nucleus (N), nuclear envelope pores (white arrows). Plasmodesmata tend to occur in rows (arrowheads). This exemplifies primary plasmodesmata that will be removed later in development (the cell is a metaxylem element; its connections will be severed before it loses its contents and dies). Another example of plasmodesmatal occlusion is in stomata, where guard cells become symplastically-isolated (the plasmodesmata that will be occluded are seen in guard mother cell walls in Plate 40). The inset is a cross-sectional view at high magnification (x700,000), showing (i) plasma membrane bilayer (Po - outer, Pi - inner lipid head group layers), (ii) desmotubule (the pale circle is the relatively unstained interior of the membrane bilayer, the central dot (Di) is the inner, tightly curved, set of head groups and Do the position of the outer head groups), and (iii) cytoplasmic annulus (CA), whose limits are hard to see because it is stained similarly to the inner head groups of the plasma membrane and the outer head groups of the desmotubule ; it contains hints of particles which may be the component that narrows the transport pathway to micro-channels.

Plate 45c, d and e Plasmodesmata are often grouped in *primary pit fields*, e.g. (c) - oat leaf mesophyll, x31,000, (d) - oblique/surface view, pea leaf mesophyll, x66,000. (e) is a shadow-cast preparation from a maize root tip (x16,000) showing cellulose microfibrils delimiting the pit area and the individual plasmodesmatal pores. In (c) the plasma membrane is closely appressed to the desmotubules at the cytoplasmic extremities of the plasmodesmata (e.g. circles). These *neck constrictions* are very common and may be part of a mechanism for regulating the diameter of the transport pathway. Membranous contortions (arrowheads) in the central part of these plasmodesmata are probably forerunners of more complex inter-plasmodesma connections (see f,g,h).

Plate 45f, g and h These three micrographs illustrate compound plasmodesmata, in which several canals, formed secondarily, meet in the interior of the wall. In (f) (between two transfer cells in the phloem parenchyma of a lupin leaf vein, x35,000, see also 46c) many canals radiate in both directions from the centre of the wall. In (g) (vascular parenchyma of *Polemonium* stem, x95,000) a lateral passage in the centre of the cell wall interconnects two plasmodesmata. In (h) the upper, empty-looking cell is a sieve element and the lower its companion cell. Many plasmodesmata funnel from the latter towards fewer canals leading into the former. On the sieve element side the canals are lined with electron-lucent callose (*Lupinus*, x35,000, see also Plates 46c, 49).

Transport through plasmodesmata is probably subject to several kinds of regulatory mechanism, apart from total occlusion. Secondary plasmodesmata in leaves can develop the capacity to transport molecules some 20-fold larger than the exclusion limit that applies to primary plasmodesmata, i.e. up to a molecular weight of about 15,000 Da. They do so under the influence of *movement proteins,* which are coded for by *systemic* viruses (able to spread from cell-to-cell in the infected plant). Plants may have their own homologues of viral movement proteins to facilitate intercellular transport of informational macromolecules in special situations, e.g. through companion cell-sieve element plasmodesmata, as described in Plate 49.

(45e reproduced by permission from *Ultrastructural Plant Cytology*, 1965, by K. Muhlethaler, Elsevier Publishing Co.; 45a and inset in (b) kindly provided by R. L. Overall, reproduced by permission from Protoplasma 111, 134, 1982).

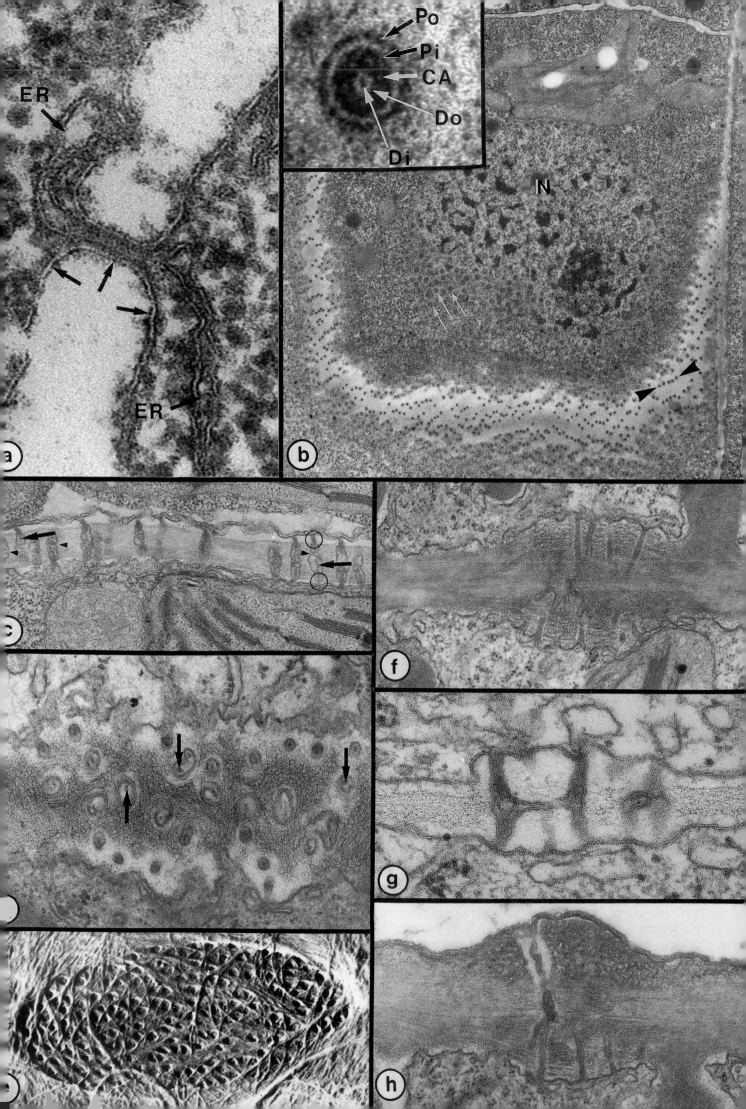

46   Transfer Cells

Introduction: Transfer cells are not a cell type in their own right. Rather they are cells of many different types which have in common a modification of the cell wall and plasma membrane. Thus there may be epidermal transfer cells, xylem or phloem parenchyma transfer cells, pericycle transfer cells, haustorial transfer cells, transfer cells in glands (etc.). The characteristic feature of transfer cells is that the surface area of the plasma membrane is augmented by ingrowths of wall material which protrude into the living contents of the cell. This may happen over the complete surface of the cell or only on particular faces of the cell. It is considered to be an adaptation which, by increasing the surface:volume ratio, enhances transport of solutes across the plasma membrane. In addition to its extra surface area, the plasma membrane may become specialised through augmentation of transport proteins such as ATPases. Although the structure is basically similar in all transfer cells, the functions can vary widely. Some are secretory (e.g. in many nectary glands) and some are absorptive, as in the three examples shown in this Plate.

Plate 46a This electron micrograph shows the characteristic "wall-membrane apparatus" of a transfer cell at high magnification. The sample was prepared by freeze-substitution and the preservation of structural details is better than in the conventionally-prepared samples shown in (b) and (c). The "wall-membrane apparatus" is an interconnected labyrinth of wall ingrowths, lined throughout by the plasma membrane (arrows). Enhancement of the surface area of the plasma membrane (PM) by the wall ingrowths is about five-fold in the area of cell surface shown here. The ingrowths are finger-like projections from the wall, branching, anastomosing and bending in all directions, often appearing in the section as apparently isolated profiles. Mitochondria are often found in the cytoplasmic enclaves that penetrate between the wall ingrowths, where they are believed to supply energy for trans-plasma membrane solute transport. In this particular sample the mitochondria (M) are not in the labyrinth but lie in the nearby cytoplasm. As is common in freeze-substituted specimens, the outer membrane of the mitochondrial envelope is scarcely stained.

This transfer cell was at the absorptive surface of a young moss sporophyte, where its haustorial "foot" was embedded in (and absorbing nutrient from) its host gametophyte generation. Transfer cells are common at junctions where one generation is nutritionally dependent upon another, in plants ranging from bryophytes to angiosperms. Indeed this situation may be where the transfer cell adaptation first arose. Kindly provided by A. Browning. *Funaria hygrometrica*, x38,000.

Plate 46b In this example, the transfer cells are xylem parenchyma cells modified by the development of wall ingrowths. Like other transfer cells, they have many mitochondria (e.g. M in cell at upper right), and considerably amplified surface areas of plasma membrane. They surround two xylem elements in a vascular bundle, sectioned where it passes out from the stem node as a leaf trace into the petiole. This part of the vascular system often displays spectacular arrays of transfer cells. Use of radioactive tracers has shown that stem nodes are vascular traffic-control regions where solutes absorbed from xylem sap pass across the intervening living cells into the phloem and are redirected to the shoot apex, and used there in growth. Parts of the xylem other than at departing leaf traces (e.g. Plate 48d) have normal xylem parenchyma cells, not modified as transfer cells.

The two xylem elements themselves display several of the features also seen in Plate 48 - microfibrillar remnants of the primary wall (arrowed circle), survival of microfibrils plus matrix in zones protected by lignin (arrows), and lignified thickenings (seen both in transverse and longitudinal section here because the xylem elements have been sectioned obliquely). Cotyledonary node of *Galium aparine*, x12,000.

Plate 46c The transfer cells in this example are associated with sieve elements in the phloem of minor veins of leaves - a site of intensive transport across the plasma membrane in many plants. Two types of solute may be transferred. One consists of products of photosynthesis, produced in mesophyll cells in the leaf. They diffuse along the cell wall network and are absorbed by phloem parenchyma cells in the veins (especially companion cells, the sister cells of sieve elements - see Plate 49). The other category of solute consists of solutes that have entered the leaf in the transpiration stream. This section shows a minor vein with four companion cells (CC) associated with two sieve elements (SE). Wall ingrowths increase the absorptive surfaces of the companion cells. The absorbed solutes are passed on to the sieve elements through plasmodesmata (one example of a compound plasmodesma is shown in the inset) for export from the leaf. Minor vein transfer cells do not occur in all species. They are mainly in herbaceous families, including the very large groups Asteraceae and Fabaceae. The transfer cell option is only one of several pathways by which products of photosynthesis are loaded into the vein phloem. Different kinds of plant exploit the various options to different degrees. *Senecio vulgaris* leaf minor vein, x6,800; inset x40,000.

The gap between the plasma membrane and the wall material in the ingrowths in (b) and (c) and especially in the inset is almost certainly an artefact caused by shrinkage of the wall material during specimen preparation. Comparison with (a), where there is no such gap, demonstrates the superiority of fast-freezing and freeze-substitution compared with chemical fixation and conventional processing.

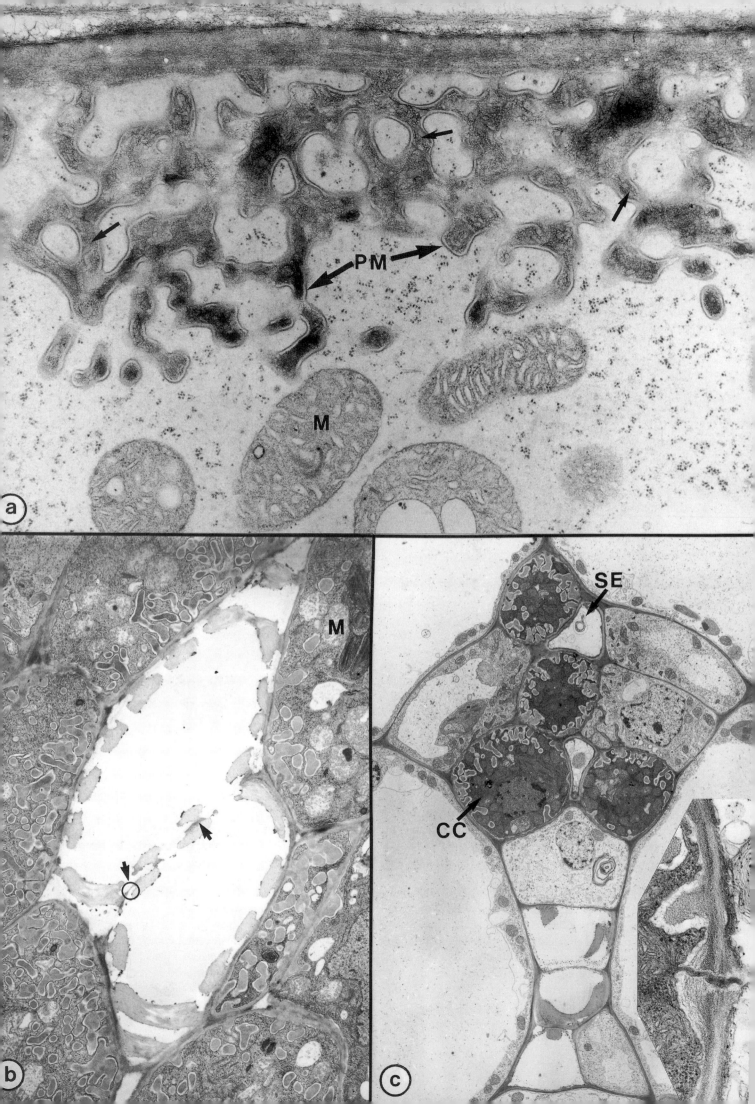

47   Xylem (1): Developing Xylem Elements

Introduction: *Xylem* tissue carries the transpiration stream from sites of water uptake (normally in roots) to sites of evaporation (in shoots, especially leaves). When mature, it consists predominantly of dead conducting tissue composed of *vessel elements* or *tracheids* aligned in longitudinal files, together with neighbouring live *xylem parenchyma* cells that transfer solutes into and out of the non-living conduits. As in phloem tissue (Plates 49,50), there may also be mechanically-supportive *fibre* cells, especially in woody plants. Developing xylem elements (not yet dead and empty) are shown in Plate 47, and mature systems in Plate 48.

Mature xylem vessels and tracheids form continuous pathways from root to stem and out through leaf traces in stem nodes to the midrib and successively finer veins of leaves. They become mechanically strengthened during their development by formation of wall thickenings, in a unique type of cell surface differentiation. Microfilaments aid in determining the ring, spiral or reticulate patterns of the thickenings, then cortical microtubules position their cellulose component, and finally lignin is deposited in their wall matrix. *Lignin* is a phenolic polymer derived by peroxidative cross-linking of aromatic monolignol units, mainly p-coumaryl, coniferyl and sinapyl alcohols. It is also cross-linked to microfibrillar and other matrix materials in the wall. Lignin-like substances occur in some advanced members of the algae but its properties have been exploited mostly in the vascular plants. It is more rigid and less permeable to water and solutes than cell walls which merely have pectic and hemicellulosic wall-matrix materials. It resists degradation.

The individual cells that form vessels and tracheids lose their contents, plasma membrane and primary wall and die at maturity, leaving a fluid filled lumen in continuity with aqueous phases in the cell wall spaces of the adjacent cells. This continuum extends throughout the plant (though it may be interrupted in places - see Plates 51 and 53) and is known as the *apoplast*. It can be viewed as a labyrinthine compartment surrounding the equally labyrinthine symplast, i.e. the living protoplasts, interconnected by plasmodesmata (Plate 45). Water and solutes in the apoplast exchange with water and solutes in the symplast throughout the plant, with perhaps the greatest fluxes at sites where the long-distance channels, xylem and phloem, are being loaded or unloaded.

Loss of water at evaporative surfaces pulls the transpiration stream, a bulk flow of water, through the xylem. Plants have exploited the existence of this basic process by adding a number of vital subsidiary mechanisms. Accumulation of mineral solutes from soil solutions into root cells and their subsequent secretion into the xylem for carriage to the shoot system is one. It serves a role in mineral nutrition of remote tissues and organs, and provides an osmotic pump, *root pressure*, that can supplement or replace evaporative pull. Nitrogenous nutrients are also carried in xylem sap, and both these and mineral salts are transferred to the phloem where the xylem and phloem transport systems are brought close together, e.g. in stem nodes and in minor veins of leaves (see also Plate 50).

Plate 47a This micrograph shows a longitudinal section of a mature *protoxylem* element (PX) and a developing *metaxylem* element (MX) in the vascular tissue of an *Azolla* root (see Plate 60 for a transverse section). Protoxylem elements mature while the surrounding tissue is still elongating, therefore their lignified thickenings become pulled apart (arrows). The black areas (asterisks) represent parts of cells that bulge into the protoxylem and so have been included in the plane of the section. Metaxylem elements mature after the tissue has elongated. Many wall thickenings are forming in this example (arrows). The end wall (W) separates one xylem element from the next in the file. Nucleus (N) and cytoplasm are still present. x4,300.

Plate 47b Part of the wall between two developing xylem elements is illustrated here in a section of *Lupinus* leaf vein. The uppermost cell still has its cytoplasm and intact tonoplast (T); Golgi stacks (G) and their associated vesicles are prominent. Cortical microtubules which overlie the developing xylem thickenings are marked by arrows. Their association with the lignified bands is very clear, and is related to cellulose deposition, not lignification. The lower cell is more advanced in its development. Its tonoplast has ruptured and only a few recognizable cytoplasmic components remain. Although its cytoplasm has been largely digested, breakdown of the primary wall has not yet commenced. All but the cellulose microfibrils will become digested from zones between the thickenings (see 47c-d). x16,500.

Plate 47c and d These light micrographs show that lignification starts in the interior of developing xylem thickenings. In 47c, pale areas (L) contrast with the darker stained non-impregnated hemicellulosic parts (H) of the young thickenings. The primary wall (PW) is still present. 47d shows a later stage, with no cytoplasm, loss of the primary wall (PW) except where protected by the lignin, and uniformly lignified thickenings. Lignified thickenings often develop back to back in neighbouring cells, as seen here, reflecting the influence of some as yet unknown form of intercellular communication, exerted during cell differentiation. *Phaseolus*, x1,800.

Plate 47e Developing xylem in transverse section (*Galium*, x13,500), showing microtubules (arrows) at the developing thickenings and many Golgi stacks (G), producing large numbers of vesicles, probably carrying wall matrix material to the developing thickenings.

Plate 47f, g and h Three stages in the dissolution of the end wall (W) of vessel elements in *Phaseolus* are seen here by light microscopy at x900. The mature vessel (derived from vessel elements) is an open pipe (47h).

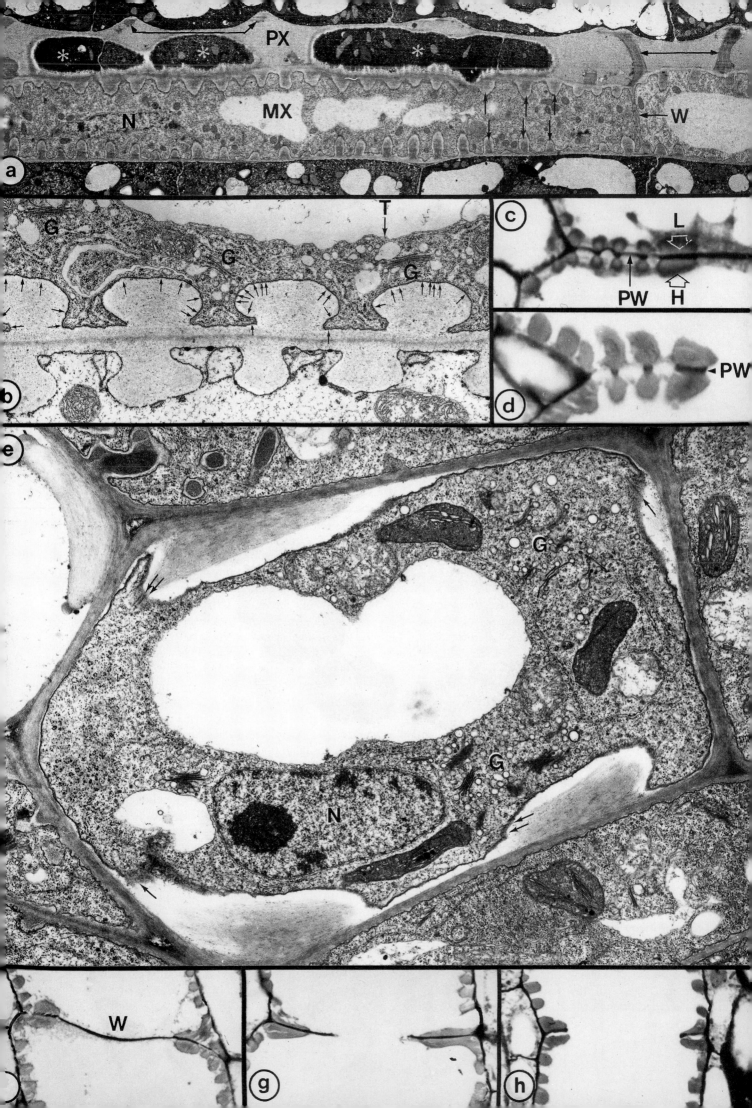

48 Xylem (2): Mature Xylem and Xylem Parenchyma

Introduction: One of the keys to the development of upright land plants was the evolution of water-conducting tissue. Tracheids occur in all vascular plants. They arise from single cells which die and empty at maturity, but retain their end walls, so files of tracheids are not fully open conduits. Vessel elements, by contrast, lose their end walls (Plate 47f-h), giving rise to open files (vessels). The constituent elements are shorter and of greater diameter than tracheids, and are restricted to flowering plants. The course of evolution is thought to have been from plants with tracheids, which served both mechanical and conducting roles, to plants with vessels, which primarily serve in conduction, together with tracheids and xylem fibres, which add mechanical support roles.

The construction of vessels from individual cells, differentiating end-to-end, is an elegant compromise in relation to their function in bulk transport of water. They can be up to 0.5mm wide and several metres long. Overlapping vessels, each a few metres long, and with pores for lateral water movement in the overlaps, provide continuity of flow through the plant and offer alternative pathways in the event of damage to any one vessel, e.g. blockage by an air bubble. The amount of water that can be carried in a pipe is proportional to the fourth power of the radius, so small increases in size can mean major enhancement of conductivity. Yet there is a limit, because if transpirational pull is to lift water up vessels from absorptive sites to evaporative sites, there must be intimate contact between the cell walls and the water columns to maintain cohesion. Also, formation of air bubbles is less likely in small channels, and if one of a large number of small channels is interrupted, the consequences are less serious than if a large one, carrying a high proportion of the total flow, were to be broken. Mature xylem is in fact a remarkably air-free tissue, thanks to special arrangements, such as the sealing properties of the Casparian strip in the endodermis of roots (Plate 51). Average flow velocities in xylem are a few mm per second, but the system is capable of more. When the transpiration stream of a wheat plant was restricted experimentally to a single vessel in a single root, the velocity of flow rose to 800mm per second.

Plate 48a Part of a long xylem element, isolated from a lettuce leaf vein by digestion with snail gut enzymes, is viewed here by scanning electron microscopy at x3,250. Except in a few places the remains of the primary wall (PW) have been digested away, and only the reticulate lignified thickenings (L) remain. Snail digestive juice contains enzymes that destroy cellulose and other wall carbohydrates. One role for the xylem of leaf veins is to replenish water lost from the evaporative surfaces of mesophyll cells. Hence it is notable that all along the sides of these xylem elements, as in (a-c), there are numerous gaps in the reticulate pattern of lignified thickenings, through which water can permeate freely.

Plate 48b In this scanning electron micrograph (x3,400) xylem elements in a pea leaf vein that was partially isolated after treatment with the enzyme pectinase are viewed (as in 48a) from the outside, though one can see into the lumen at the broken end of the topmost element (white arrow), where both outside and inside faces of the lignified thickenings are visible. The continuous wall between the thickenings has suffered tearing (asterisks) during specimen preparation. It is probably composed largely of cellulosic remnants of the primary wall, and is retained here more than in 48a because of the absence of cellulose-digesting enzymes during preparation.

Plate 48c In this ultra-thin section (x7,200) of part of two xylem elements in a pea leaf vein, living xylem parenchyma cells just enter the picture at top and bottom. The lignified thickenings and the wisps of cellulose microfibrils (arrows) that interconnect them are the only components to survive the auto-digestion processes of xylem maturation. These wisps presumably constitute the expanses of continuous wall seen at much lower resolution in Plate 48b. The thickenings of the upper element are much more widely spaced than those in the lower, probably indicating that it matured earlier, and became passively stretched, as in protoxylem (Plate 47a).

Plate 48d This transverse section of primary xylem tissue in the stem of a fumitory (*Fumaria*) seedling includes the following features of xylem elements (X): lignified thickenings, remnants of primary wall in exposed localities (arrowed circles) and more complete survival in areas *underlying* lignified thickenings (arrows). An important feature not previously illustrated is that living xylem parenchyma cells lie around and between the xylem elements. They are responsible for loading the xylem sap with solutes such as mineral nutrients, amino acids, and some hormones, and also for absorbing solutes from the xylem sap. Roots are sites of loading; stem xylem parenchyma may both load and unload the xylem. The xylem parenchyma of stem nodes, where vascular bundles pass out into leaf petioles, is particularly active in this regard, and the cells are often modified to become transfer cells (Plate 47). Note the many cisternae of endoplasmic reticulum in the xylem parenchyma cells. Such cells can be shown to absorb amino acids from the xylem sap and to incorporate them into protein. x2,700.

The inset at lower left shows a scanning electron micrograph of a primary xylem element (comparable to those in the section) with helical lignified thickenings (x2,200). Helical and annular thickenings do not prevent cell extension. The inset at upper right shows secondary xylem, where the lignification was laid down after the cell had expanded and is much more extensive, covering the whole wall except at lens-shaped pits (x2,200). Both insets are of castor bean (*Ricinus*) stem, and were kindly provided by J. Sprent and J. Milburn.

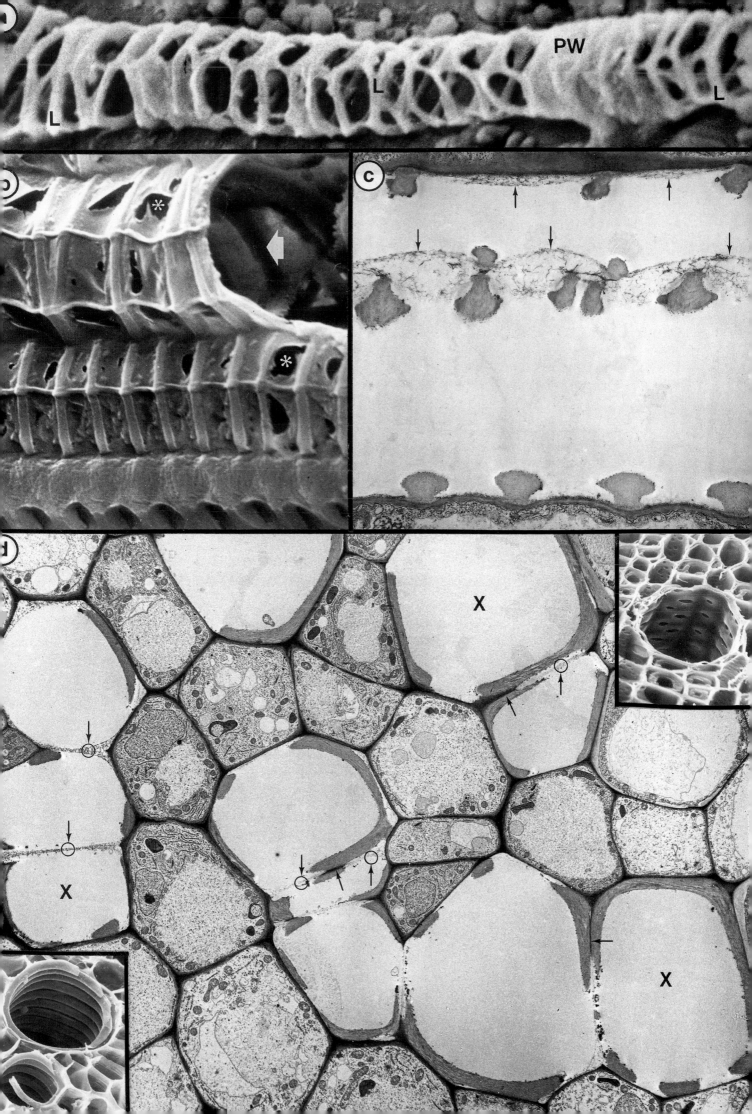

49 Phloem (1): Sieve Element and Companion Cell

Introduction: Plasmodesmatal connections provide a route for *short distance transport* of low molecular weight solutes within symplastic domains, i.e. groups of cells all bounded by a common plasma membrane (Plate 45). Demand for *long distance transport* of solutes around the whole plant between sources and sinks is met by the *phloem* component of the vascular tissue. *Sieve elements* and *companion cells*, illustrated here, are the major cell types active in phloem transport. The two lie alongside each other, and in fact are products of division of common parental cells. Sieve elements grow into long cylindrical cells, arranged end-to-end to form *sieve tubes*. Plasmodesmata in the end walls of young sieve elements become enlarged during cell maturation into plasma membrane-lined *sieve plate pores,* thus enhancing the capacity of the sieve tubes to function in bulk long-distance transport. Companion cells function in loading and unloading sieve elements.

The companion cell/sieve element/sieve tube continuum is a symplastic compartment, lined by a common plasma membrane that passes through companion cell - sieve element plasmodesmata and sieve plate pores. The continuum functions as an extended osmotic pump, delivering sucrose (or other soluble carbohydrates) from *sources* (in photosynthesising leaves, reserves in germinating seedlings and sprouting storage organs) to *sinks* where substrate is required (growing stem and root tips, expanding leaves, reproductive structures, growing seeds and storage organs). The sucrose (etc) is accumulated in source sites, which thereby develop a high osmotic pressure. Removal of sugars at sinks leads to a pressure difference between source and sink that drives a mass flow of sugar solution through the sieve tubes from the one to the other.

The structure and function of sieve elements is hard to investigate. The high positive pressure of their contents, and their ability to invoke rapid wound responses, can alter their structure and stop their normal function within milliseconds of the onset of manipulations. This is tantalising, for they are one of the most remarkable cell types in the plant kingdom. In angiosperms they have no nucleus when they are mature - one of the few features that distinguishes flowering plants from groups of lower plants. Nevertheless they can survive for decades, even centuries, in long-lived monocotyledons such as palm trees. They retain mitochondria, non-photosynthetic plastids, plasma membrane and a form of endoplasmic reticulum, all of which must presumably be maintained over long periods, but no ribosomes, tonoplast or vacuole can be recognised.

It seems that companion cells do much more than load sieve elements with solutes to be transported. There is evidence that they also provide sieve elements with essential proteins, transcribed from genes in the companion cell and somehow targeted into the sieve element. In accord with this role, plasmodesmata between companion cells and sieve elements have been found to have unusually high molecular weight exclusion limits and can pass (at least) small proteins. This shows that plants can regulate plasmodesmatal permeability, but it is not clear how they do so. Systemic plant viruses can facilitate their own spread from cell to cell by causing synthesis of movement proteins, which become built in to secondary plasmodesmata and allow large molecules to pass through (see Plate 45). It may be that plants have the necessary genetic information to make their own proteins for modifying plasmodesmatal permeability, and use them in situations such as the sieve element - companion cell junction. This genetic information may in fact be where systemic viruses acquired their genes for movement proteins. Movement proteins may work by dilating the lumen of plasmodesmata and/or by altering the conformation of macromolecules so that they are more readily transported (possibly limited to proteins that possess an appropriate signal sequence).

The sieve element (lower) and companion cell (upper) shown here in cross section are representative of the two principal cell types of phloem tissue.

Sieve elements mature following the breakdown of their tonoplast and much of their cytoplasm. In contrast to xylem cells (Plates 47 and 48) the plasma membrane of the sieve tube (PM) remains intact, in fact the osmotic transport system depends upon survival of its semipermeability. The lumen of the sieve element contains mitochondria (M) and plastids (P). One of the most abundant proteins of sieve elements is known as *P-protein* (arrows, see also Plate 50).

The relative size of sieve elements and their sibling companion cells differs according to position in the long-distance transport system. Where phloem loading predominates over longitudinal movement, as in minor veins of leaves, which accumulate photosynthates from the mesophyll tissue, the companion cells have a greater diameter than the sieve elements. They may also show surface area amplification and develop a large population of mitochondria, both features related to the demands of energy-requiring solute accumulation (see Plate 45c). Conversely, in stems, where longitudinal movement between sources and sinks predominates over retrieval of leaked solutes, the sieve elements are the larger. The present example, part of the phloem in an ovary of the grape hyacinth, *Muscari,* is intermediate. x3,000.

Sieve element-companion cell plasmodesmata are shown in the wall between the two cell types (PD) (see also Plate 45). Linings of electron transparent callose (C) are typical of plasmodesmata in this situation, as also in sieve plate pores (Plate 50). Callose deposition can occur very rapidly, indeed some of the callose seen in the micrograph probably appeared during specimen preparation as a form of wound reaction.

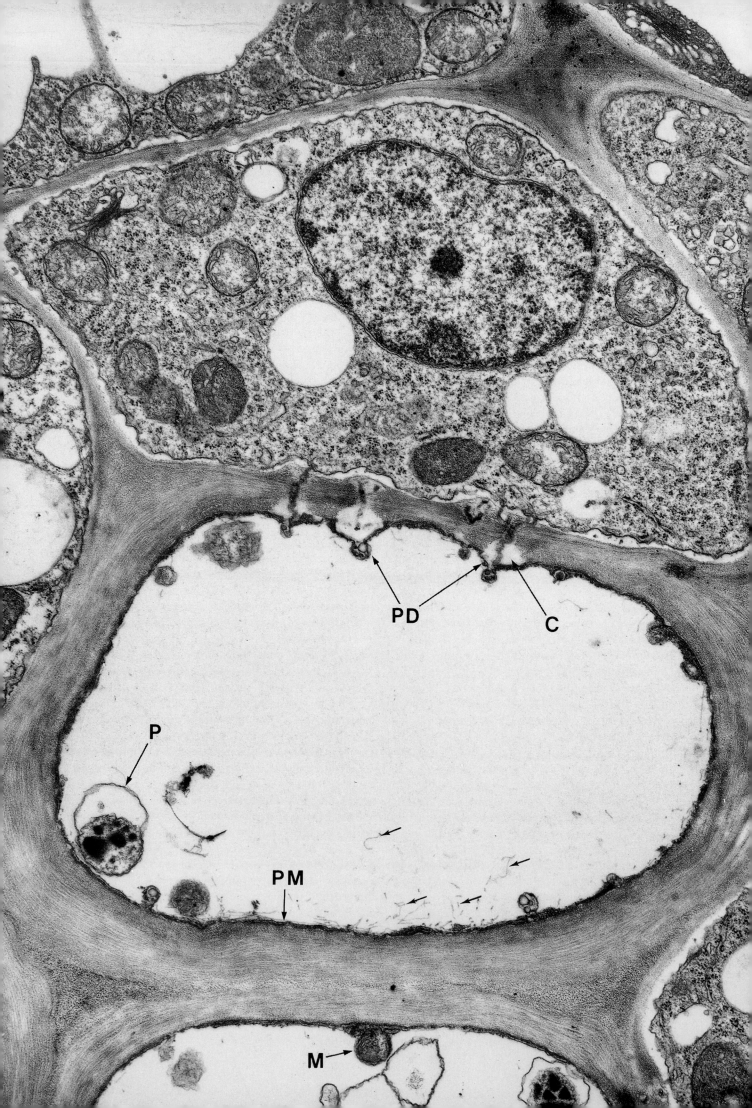

50   Phloem (2): Sieve Plates and Sieve Plate Pores

Introduction: This Plate complements Plate 49 by showing details of sieve plates and sieve plate pores.

Plate 50a In longitudinal section sieve tubes (ST) are seen to be interrupted at intervals by perforated cross walls, the sieve plates (SP), which were originally the walls between successive cells in the file of differentiating sieve elements. Sieve plate pores develop from plasmodesmata in the primary cross walls. In this particular tissue some of the accompanying companion cells have become specialized by the formation of wall ingrowths (see also Plate 46). Note the abundance of compound plasmodesmata between companion cells and sieve tubes (arrowheads), pathways through which the sieve tubes are loaded or unloaded. This sample is from a petiole, an organ in which the emphasis is on longitudinal transport rather than loading or unloading. However, there is probably some lateral exchange at all points in the vascular system. In particular, solutes that are being carried in the transpiration stream in the xylem are passed across the intervening cells to the phloem, especially in stem nodes, for onward transport to sinks. For example, transport of amino acids and amides from xylem to phloem delivers much of the nitrogenous input into developing fruits, and provides growing stem apices with a balanced diet of carbon and nitrogen compounds.

The sieve elements lumen contains a dispersed network of P-protein. At higher magnifications the plasma membrane and sparse population of mitochondria and plastids are more obvious. P-protein fibrils arise in developing sieve elements as aggregates (termed slime bodies in earlier literature), then individual fibrils disperse throughout the sieve tube when the tonoplast breaks down and the normal cytoplasmic ground substance is lost. White lupin petiole, x3,500.

Plate 50b Sieve plates and their pores are exceptionally difficult to visualise in their native state because wound reaction mechanisms alter their appearance almost as soon as a tissue sample is taken from the plant. These changes can continue during the early part of the subsequent fixation period. The only ways to minimise them are to freeze intact material as quickly as possible (which can never be fully satisfactory because the phloem tissue is always internal and therefore cannot be cooled as rapidly as surface-located cells), or to use chemical fixation on intact plants (as distinct from excised tissue samples) without subjecting them to any mechanical damage. These wound reaction mechanisms probably evolved to prevent prolonged osmotically-driven leakage of sugar solution from the sieve tubes after mechanical or insect damage to the plant.

This micrograph shows a sieve plate in a condition thought to be close to the normal state. Only a very small amount of callose (C, the electron-lucent ring around each pore) is present (cf. (d)) and the pores contain relatively few fibrils of P-protein (P). Since sieve tubes are under considerable hydrostatic pressure, due to the osmotic effect of their sugary content, damage leads to a sudden release of pressure and a surge of liquid over long distances down the sieve tubes. Masses of P-protein fibrils (normally uniformly dispersed throughout the sieve elements) surge onto the sieve plates and into the pores. A temporary blockage of the pores is thus effected while another reaction takes place, the formation of massive callose plugs which seal the pores (d). Callose formation is extremely rapid and the response system is sensitive to chemical as well as to mechanical damage to the phloem. Some insects, notably aphids, do manage to circumvent these response systems and feed on phloem sap by inserting their stylets through the surrounding tissue and the sieve element wall. They create only a small pressure drop in the parts of the sieve tubes tapped by their stylets, so they can successfully rob the plant of its food supplies.

A specialized form of endoplasmic reticulum, consisting of layers of flat, smooth cisternae, is found in sieve elements (ER). The cells to left and right belong to companion cells. Same tissue as in 50a, x27,000.

Plate 50c This scanning electron micrograph of a sieve plate in a water lily petiole was obtained by transfusing the whole petiole with glutaraldehyde in an attempt to avoid sudden pressure losses in the sieve tubes. The tissue was then critically point dried to maintain structural arrangements as they were at the time of fixation. Finally, individual sieve tubes and plates were exposed by dissecting away the adjacent cells. Sieve plate pores are visible (arrows). Strands of filamentous P-protein are much more obvious in this mode of preparation than in ultra-thin sections. Petiole of *Nymphoides peltata*, x12,000. Micrograph kindly provided by R. Johnson.

The structure of the sieve plate, and the amount of P-protein in its pores, are key factors in considering the capacity for translocation in the phloem. At the plates, the available pathway for movement is not the lumen of the sieve tube, but the sum of the lumina of a set of very much smaller pores, each one occupied by fibrillar protein to an extent that is difficult to quantify. Clearly, flow of sugar solution along sieve tubes is constrained at the sieve plates more than elsewhere. Countering this disadvantage, the plates provide frequent sites along the vascular system at which the flow can be plugged in the event of damage. A further function, as sites at which a motive force for phloem translocation might be applied to the sieve tube contents, has been suggested.

Plate 50d This face view of part of a sieve plate, seen in an ultra-thin section, shows the appearance of pores after the system has reponded to damage. Each one has a copious deposit of electron-transparent callose constricting the original lumen and squeezing the P-protein fibrils that had surged into the pore. The weft of cell wall microfibrils that delimit the pores in the sieve plate can be seen. *Coleus blumei* stem, x19,000.

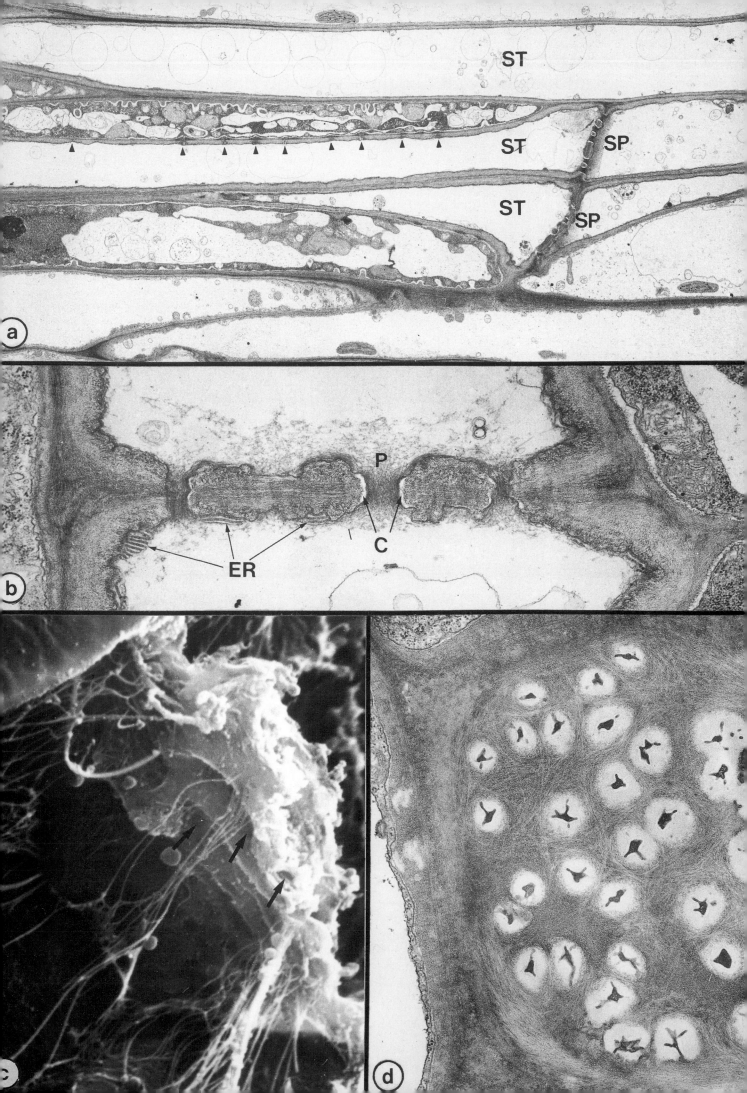

51 Endodermis and Casparian Strip

Introduction: Most cell walls are sufficiently porous (Plate 4) and hydrated to allow water and low-molecular weight solutes to pass through, although negative electrical charges in anionic wall matrix polymers may impede movement of cations. The upper size limit for diffusion through walls depends on the pore dimensions, which vary widely according to the nature of the wall. Thus mucilage molecules with molecular weights measured in the millions can be extruded through root cap cell walls (Plate 12). At the other end of the spectrum, lignified walls have very low porosity (Plate 47), even for water molecules. Many proteins are secreted *into* cell walls, and, given sufficient time, small proteins can penetrate *through* root epidermal walls or walls of cells in tissue culture. The process is, however, slow, and may not have biological significance, except in special situations, for example in an incompatibility mechanism involving penetration of ribonuclease protein molecules secreted by stylar tissue into pollen tubes (Plate 57). As a rough generalisation, molecules larger than about 6,000 Daltons need long times and/or high concentration gradients to diffuse through "normal" walls to a significant extent. By contrast, the cell wall compartment is an extensively used avenue for transport of smaller molecules over short distances in plant tissues (Plate 47). An example is movement of products of photosynthesis from mesophyll tissue to leaf veins (Plate 46).

Just as the symplast does not extend without break throughout the plant (Plate 45), so too the apoplast is compartmentalised. Here we show a specialised wall-membrane apparatus that creates a localised seal in the apoplast. The porosity of the wall is reduced by impregnation with an impervious matrix material, *suberin* (see Plate 52 for composition), and the plasma membrane is locally attached to the *suberised* wall to complete the seal. The seal is analogous to intercellular *tight junctions* that prevent (for example) intestinal fluids from leaking between cells lining the gut into the body tissues of animals. Here the seal is the *Casparian strip* in the *endodermis* cell layer of roots. Other examples of apoplastic compartments delimited by cell wall impregnation are shown in Plates 24, 53 and 55.

Large volumes of water move laterally across the cortex of roots *en route* to the stele for transport to the shoot in the xylem. Some goes through cells and some through cell walls. At the same time, solutes are accumulated selectively from the soil solution into the surface and cortical cells of the root. Plasmodesmata provide a symplastic pathway for these to move to the interior of the root, where they are secreted into the xylem sap. Accumulation at the outer end of the pathway, and secretion from the inner, xylem parenchyma, cells combine to create a concentration gradient, down which the solutes diffuse from cell to cell. Once in the xylem, a non-living space, solutes could leak away by diffusion through the apoplast, thus dissipating their osmotic effect. However, they do not do so, because the apoplast around the vascular tissue is sealed by the suberised Casparian strip in the radial walls of the endodermal cell layer. This seal also constrains the inward flux of both water and solutes to endodermal cell plasma membrane and plasmodesmata. Once the solutes have been secreted into the sealed-off stelar apoplast, their osmotic effect draws water from the cortex into the stele and from there up the xylem, i.e. the existence of the Casparian strip underlies the phenomenon of root pressure (Plate 47).

Plate 51a Arrowheads point to the Casparian strips in the endodermis cell layer in this transverse section of the stele of an *Azolla* root. The strips stain differentially because of their suberin content. They pass right around all radial walls of each endodermal cell. The cell types are labelled for comparison with Plate 60, which shows the same types in the same spatial arrangement in an immature part of the root. E, endodermal cell; P, pericycle cell; S, sieve element; PP, phloem parenchyma; PX, protoxylem; MX, metaxylem. x1,700.

Plate 51b Legume root nodules import carbohydrates *via* the phloem to provide an energy source for nitrogen-fixing bacteria in the nodule, and carbon skeletons for the assembly of fixed nitrogen into amino acids and amides. The latter are secreted, by transfer cells, into the apoplast of the nodule vascular bundle, where they are contained by the encircling endodermis and Casparian strips. The presence of high solute concentrations inside the endodermal sheath promotes osmotic uptake of water into the vein, which forces amino acid-rich solution up the xylem and into the transpiration stream of the main root. This micrograph shows a radial wall between two endodermal cells (E). Part of a cortex cell is seen at the left, and part of a pericycle transfer cell at the right. The Casparian strip lies between the open arrows, distinguished from the neighbouring wall by its more homogeneous texture, lack of middle lamella (ML), and smooth, adherent plasma membrane. Sweet pea (*Lathyrus*) root nodule, x15,000.

Plate 51c A portion of Casparian strip (top) is contrasted here with a portion of normal cell wall (bottom). The smooth texture of the strip is due to impregnation of all of the spaces between the microfibrils with suberin. The plasma membrane (PM) shows its dark-light-dark bimolecular layer appearance against both types of wall, but very conspicuously so at the Casparian strip because it is unusually flat there. The plasma membrane sticks tenaciously to the Casparian strip if endodermal cells are plasmolysed, demonstrating the efficacy of the seal.

The compartment across the middle of the picture is a flattened vacuole (V), bounded by its tonoplast (T). The tonoplast and plasma membrane are equally thick, but the dark-light-dark layers of the endoplasmic reticulum (ER) are thinner (circle). Material as in 51b. x120,000.

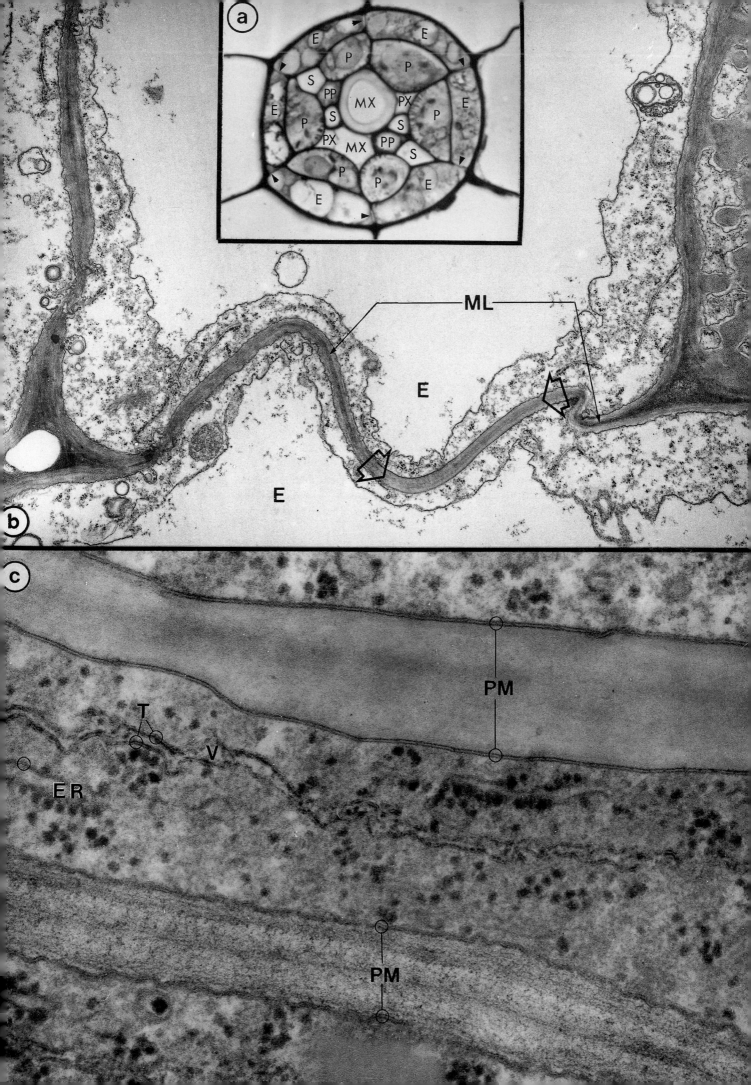

52 Wax and Cuticle

Introduction: Above-ground parts of plants are sheathed by layers of wall substances that protect against pathogens and limit loss of water. These substances, *cutin*, *suberin* and *waxes*, were probably crucial elements in colonisation of the land, when plants first met the problems of terrestrial life. They are synthesised in epidermal cells and secreted to their outer surface, a surface that is established in the single-celled embryo and thereafter is extended to cover all parts of the plant. The outer epidermal wall is augmented by microfibrillar and wall matrix materials to keep pace with its extension, and the specialised surface components are also added, usually in different forms in roots, stems and leaves, and usually in pliable forms during the growth phase, becoming more rigid after growth has ceased.

Waxes on leaf and stem surfaces are mixtures of hydrocarbon molecules, predominantly paraffins and related alcohols, ketones, acids and esters. The carbon chain lengths, in the range 20-35, are longer than in membrane lipids, and mostly long enough to ensure that they form solids at normal temperatures. As illustrated here, they often crystallise into rods, granules, threads, scales, or more amorphous crusts. They usually lie as a waterproofing layer on the outermost surface, surmounting a layer of cuticle but not chemically linked to it. They can be dissolved away with organic solvents, and can then be shown to consist of complex mixtures. The components are made in the epidermal cells and are extruded through the outer walls to the exterior, where they take up their final form. Wax patterns and crystal morphology are related to the chemistry of the compounds present. They are usually distinctive, not just of the plant species, but often of cell types (e.g. stomata differ from epidermal cells) and stage of development.

Cutin and suberin, by contrast, are polymeric compounds, not readily dissolved by simple solvents. Cutin is the material of cuticles. Suberin impregnates the wall matrix compartment of cork, bark, suberised lamellae within walls (e.g. Plate 24) and Casparian bands (Plate 51). It is also deposited after tissue wounding and as a defence layer around invading pathogens. In both cutin and suberin the building blocks are families of 16 and 18 carbon fatty acids, which are modified by hydroxylation and epoxidation. These units become extensively cross-linked by ester bonds to form non-crystalline polyesters. Suberin also contains phenolic compounds, and therefore stains differently from cutin. As in the case of waxes, cuticles are assembled into their final form after the constituents have been extruded through the outer wall of epidermal cells. Light and exposure to the atmosphere may play a part in their polymerisation at the outer surface. In most cases waxes have to be extruded before the cuticle matures, but some cuticles retain channels or pores through which substances can pass.

This Plate illustrates some of the range of form exhibited by waxes and cuticles, and also shows that these layers are not confined to the exterior of the plant. Cuticular layers can develop on the surface of internal cells, for example in sub-stomatal cavities. Some intercellular spaces are lined by hydrophobic substances, ensuring that they remain air-filled and do not become waterlogged.

Plate 9a and b Wax formations on the lower surface of a banana leaf (scanning electron microscopy; 9b, x320; 9a shows the central region of 9b magnified to x1,500). Curled threads of wax completely clothe the epidermal cells, except for the outer cuticularized caps of the stomatal guard cells around the slit-shaped stomatal apertures. The extruded wax threads are 25-50μm long, and represent just one of many possible morphologies assumed by leaf surface wax. Kindly provided by P.J.Holloway and E.A.Baker.

Plate 9c and d Ridged cuticles on epidermal cells, viewed by ultra-thin sectioning (9c) and scanning electron microscopy (9d). The ultra-thin section in 9c shows the cuticle layer (C) external to the microfibrillar wall (W). The 3-dimensional shape of the ridges is better displayed in 9d, at upper right and upper left. A smooth cuticularized cover of a pair of guard cells is the major feature of 9d, around the central stomatal aperture. The ability of individual cell types to regulate the form of their adcrusting wax layers is highlighted by comparing the specific patterns of wax distribution and cuticle morphology shown by guard cells and epidermal cells in 9a-d. 9c, daffodil flower corona, x22,000; 9d, *Auricula* sepal, x3,000.

Plate 9e and f This section (9e, part of a hair from the dead-nettle, *Lamium,* x15,000) and scanning electron micrograph (9f, a hook-shaped clinging hair from the goosegrass, *Galium aparine,* x1,000) show another common type of cuticle morphology, in which the outer surface is thrown into warty projections. Plant hairs frequently display reinforced cell walls, covered in cuticle or even silicified. Longitudinally-oriented endoplasmic reticulum cisternae in (e) probably reflect longitudinally-directed cytoplasmic streaming, often seen in plant hairs.

Plate 9g The wall seen here is lining one angle of an intercellular space (IS) in the floral nectary of *Vicia faba*. It is heavily cuticularized. Since these cells secrete nectar (an aqueous solution of sugars, in this case sucrose, glucose and fructose), it is interesting to note the presence of very numerous branched channels leading from the microfibrillar part of the wall (W) through the cuticle layer (C) towards the intercellular space where the nectar accumulates. In this nectary the secreted sugar solution emerges from intercellular spaces to the exterior *via* modified stomata which are unable to regulate their pore dimensions, unlike foliar stomata. Magnification x16,500. Micrograph kindly provided by R. Brightwell.

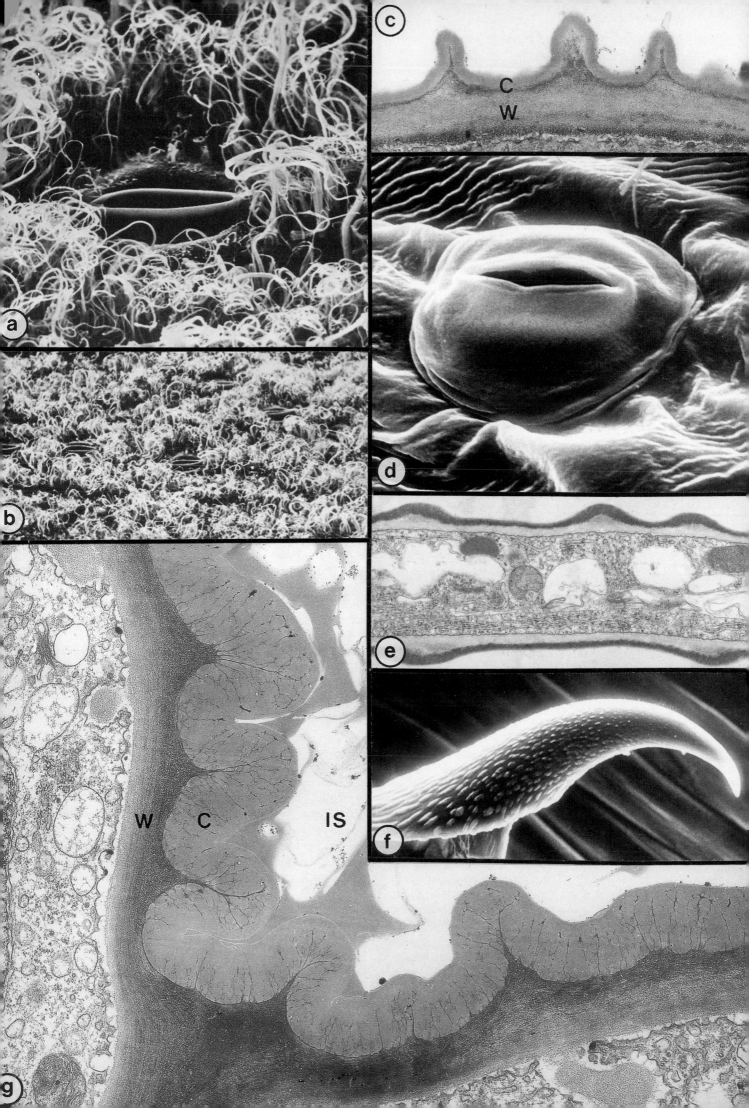

53   Apoplastic Barriers in Glands

Introduction: Plants develop many kinds of glands or secretory tissues. They may be internal, as in resin ducts, or external, as in two examples here. Secretory products, and hence ultrastructural details, vary widely (compare Plates 8,10,12), like the general anatomy of the glandular tissue. The cuticle over gland cells is commonly modified in relation to export of the product. Also the walls of the subtending cells of glands are equipped with apoplastic seals analogous to Casparian bands.

Plate 53a The morphology of this gland, on a young leaf of *Lamium*, is typical of many. The secretory cells surmount a stalk cell (S), which itself sits on an extension from a modified epidermal cell. Raw materials enter the glands *via* the numerous plasmodesmata on the walls of the subtending cell and the stalk cell (arrows). The secretory product passes out through the inner, microfibrillar zone of the cell wall of the secretory cells (W) and accumulates in the sub-cuticular space prior to its final exit, which may be through specifically located pores (not seen here) or by a more general but slower percolation. The cuticle (C) commonly detaches from this underlying wall (asterisks). x3,300.

Plate 53b The cuticle over the gland is continuous with that of the epidermal cells (CE in (a)). A special zone in the side walls of the stalk cell (stars in (a)) creates an apoplastic seal between the gland and the subtending tissue. Here the microfibrillar wall outside the plasma membrane is impregnated with an impervious matrix that prevents back-leakage of secreted products into the subtending epidermis, and restricts loss of water from the leaf. Whether the impregnation is of cutin or suberin is not clear in this case. Two groups of microtubules (arrows) lie at the extremities of the impregnated zone. They might, earlier in the development of the gland, have participated in specifying the distribution of cytoplasmic components that were responsible for synthesizing and/or depositing the impregnation. x22,000.

Plate 53c This micrograph shows the cuticle and the underlying microfibrillar layer over part of one secretory cell. The space between the two is narrow here, but it can distend markedly during secretion. As is common in secretory and absorptive cells, cisternae of cortical endoplasmic reticulum (ER) approach very close to the plasma membrane (PM) (arrows, see also Plates 8, 16, 36). This close juxtaposition may well be related to the transport of solutes between internal compartments of the cell and the extra-cytoplasmic spaces. In other gland cells, marked changes in the disposition of endoplasmic reticulum cisternae occur when secretion is initiated or ceases. Golgi stacks (G) are also present. x51,000.

Plate 53d-m: These micrographs illustrate glandular hairs (trichomes) that secrete nectar, in this case a solution of sucrose, glucose and fructose, in the nectary of *Abutilon* flowers. The system illustrates a symplastic pathway operating in concert with an apoplastic barrier.

Plate 53 d,e,f: Scanning electron micrographs (x8, x35, x500,) to show (d) the inner surfaces of the sepals with the flower parts removed to reveal the five nectaries. Each nectary is a group (e) of hairs (f). The hairs are single files of cells in their upper region and in the stalk, but the tiers just above the stalk are multicellular. The apical cell, through which the nectar is released, is round.

Plate 53 g,h: Nectary hairs (g) and hair bases (h) are seen here in sections. (g) shows the epidermal cells (E) from which the hairs emerge, the single stalk cells (arrows), and the rounded apical cells, each with a thin cuticle that has "ballooned" out from the cell itself (arrowheads). In *Abutilon* nectaries the secretion process is pressure driven. Nectar accumulates under the cuticle and when sufficient pressure has been built up, a minute cuticular pore opens like a release valve, allowing a droplet of sugar solution to be ejected. The volume of secretion from each hair per unit time can be calculated by observing these discrete pulses of secretion. (h) shows that fluorescent compounds, probably phenolic, occur in the walls at the basal region of the hairs, possibly local deposits of a form of suberin. The deposits do not extend more than a few cells up each hair. (g) x750, (h) x1,000.

Plate 53i,j,k: A thick slice of nectary tissue (i) was placed in calcofluor, a fluorescent dye that stains β-linked polysaccharides. (j) shows that the dye does not penetrate through the cuticle of the hairs. It can pass through the apoplast in the *underlying* tissue, but not past the bases of the hairs. If the dye is presented to a hair that has been broken off *above* its basal region, it penetrates along the hair (k). The experiment indicates that there is an apoplastic barrier at the hair base. (i,j) x280, (k) x350.

Plate 53l: This electron micrograph corresponds to part of (h). Electron-dense deposits impregnate the wall where the fluorescence is observed (arrows). Also a narrow deposit of material that resembles cutin (when viewed at higher magnification) (small arrows) lies between the plasma membrane and the side wall of the stalk cell. The plasma membrane adheres to this layer, just as it does at the Casparian band in endodermal cells (Plate 51). The apoplastic seal in *Abutilon* may be a combination of wall impregnation and plasma membrane sealant. x6,200.

Plate 53m: The apoplastic seal is believed to constrain the flow of pre-nectar to the cytoplasm of the stalk cell, *en route* to the apical cell. The distal cross wall of the stalk cell has about 12 plasmodesmata per μm^2, providing a symplastic pathway into the upper part of the hair. A small area of this wall, with some of its plasmodesmata, is shown here in face view. Calculations based on the dimensions of the cytoplasmic annulus (see Plate 45) show that these plasmodesmata can carry the observed flow rate of nectar, provided that there is a quite small pressure to drive the flow. x70,000.

(e-l) reproduced by permission from *Aust. J. Plant Physiol.* **3**, 619-637, 1976.

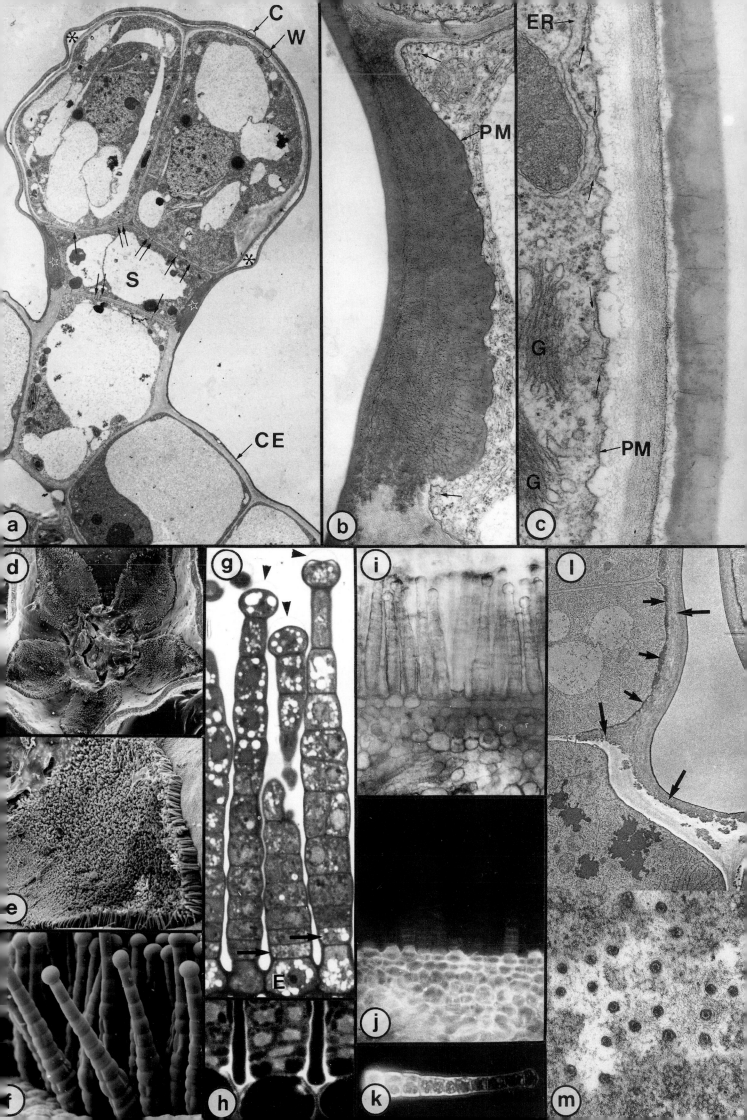

54 Pollen Grains (1): Developmental Stages

Introduction: Sexual reproduction in flowering plants involves unique developmental sequences and cell structures. The parental, flower-bearing plants are diploid *sporophytes*. *Spore mother cells* in specialised tissues undergo meiosis to produce haploid male spores (in anthers) and female spores (in ovules). The two separate gamete-bearing *gametophyte* generations (male - *microgametophyte*, female - *megagametophyte*) develop from these spores. On the female side, formation of *megaspores* and their development into megagametophytes (*embryo sacs*) is illustrated in Plate 59.

Development of male reproductive cells in representative higher plants is described here and in the following four plates. The sequence begins with meiotic formation of haploid *microspores*, which develop into *pollen grains* (the microgametophyte generation). Pollen is specialised in relation to transfer of male DNA to the female gametophyte to effect fertilisation of the egg cell, usually from one parent plant to another (cross-fertilisation). The actual male gametes (*sperm cells*) are formed before pollination in some plants and after it in others. Nutrition and gamete motility in lower plants have been touched on briefly in Plates 32a and 46a.

The products of meiosis, the newly formed microspores, lie in an isotonic fluid contained within the anther loculus, isolated from the symplast of the parent. This loculus is lined by a specialised nutritive tissue, the *tapetum*. Haploid microspores undergo just two cell divisions, the first producing a *vegetative cell* enclosing a *generative cell*. The latter re-divides to give rise to two *male sperm cells*, either before the pollen germinates or in the pollen tube after germination. Meanwhile the pollen grains develop a wall that will enable their contents to survive the extreme desiccation hazard of release into the atmosphere - a hazard maximised by their unfavourable ratio of surface area to volume. This is followed by accumulation of food reserves, starch grains or oil droplets and in some cases cell wall precursors, for the subsequent growth of the pollen tube.

Plate 54a In a very early stage of pollen development, the pollen (microspore) mother cells of the sporogenous tissue of the anther become isolated by deposits of callose (C) so massive that the primary walls are obliterated, apart from the middle lamella (circles). As in sieve plates and plasmodesmata (Plates 49, 50 and 45h) the callose appears virtually structureless and electron-transparent. Initially the pollen mother cells are interconnected by cytoplasmic bridges (PD*), facilitating rapid distribution of nutrients and synchronizing the development of the

cells. Remnants of true plasmodesmata (PD) persist. Both types of intercellular connection are eventually sealed by callose deposition. *Avena sativa,* x22,000.

Plate 54b and c These micrographs show microspores of *Avena sativa* at a later stage of development. The callose walls have been digested away (see Plate 55), perhaps providing substrate for development of the mature walls, which have now been laid down. These microspores have not long been released from the tetrads formed at meiosis. They are vacuolated and have not yet undergone division or accumulated food reserves. 54b is an enlarged (x16,500) detail from 54c (x2,800). In 54c, maturing pollen grains are seen in the anther loculus (L), the wall of which consists of three layers of cells (stars) (the innermost layer partially crushed). The loculus is lined by the tapetum (T) (for details see Plate 55).

The pollen grain wall is made up of a thin inner layer called the *intine* (I), of microfibrils and matrix, and the outer *exine*, composed of *sporopollenin*. Sporopollenin is one of the most resistant biological polymers known, being unaffected by enzymic decay and attack by strong acids. Like lignin, it is a group of related polymers rather than a single substance of defined composition (see Plate 56). The exine is subdivided into an inner platform, the *nexine* (N) from which rods, or *bacula* (B), extend like columns. A roof, or *tectum* (T) surmounts the bacula. Oat pollen grains, like various others, have a single, circular pore (see 54b), where the exine bulges in a rim around a separate lid, or *operculum* (O), of sporopollenin sitting in a thickened pad of intine. The pore arises on that face of each grain that lies outermost in the tetrad. It is the site of pollen tube emergence during germination of the grain, and it may also function in water economy, opening in moist conditions and remaining shut in dry conditions. The pad of wall below the pore and the intine in general are major sites of hayfever antigens that diffuse rapidly out of moistened, mature pollen grains.

Tapetal cell nuclei are typical of many cells of very specialized function in being highly heterochromatic. The remaining euchromatin may control specialized tapetal functions (for further information see Plate 55), one of which is visible here and involves production of a carpet of *orbicules* (arrows, 54b). These irregularly spherical bodies become coated with sporopollenin and line the anther loculus. They resemble units of the pollen grain wall, but there is evidence that the latter is assembled *in situ,* and *not* by transfer of orbicules from the tapetum as "building blocks".

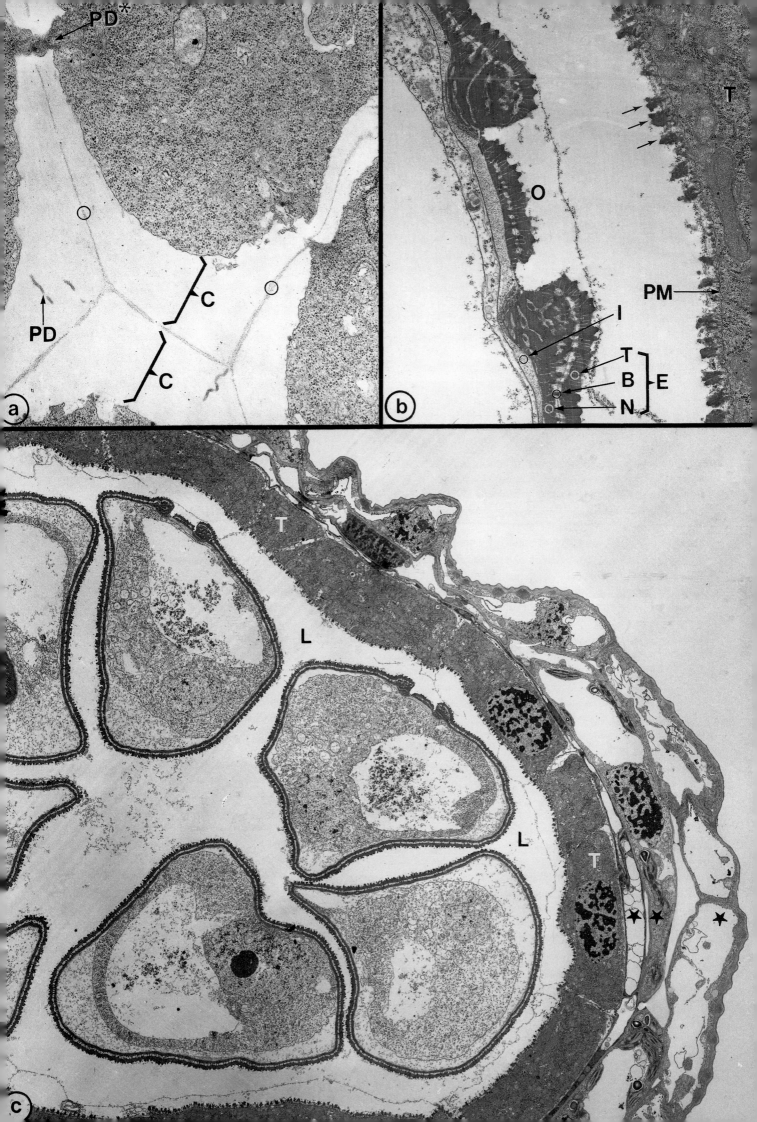

55   The Cytoplasm of Tapetal Cells

Introduction: Tapetal cells and microspores are dramatic examples of differential gene expression, for although they are strikingly different they are both derived from the same parental cells in the developing anther. Each final division cuts off a future tapetal cell from a future microspore mother cell. The microspore mother cell undergoes meiosis and the tapetal cell carries out mitosis, but without cytokinesis, so that a binucleate cell is formed. Binucleate conditions occur in some other specialised secretory cells, for example in animal liver. Tapetal cells then lose their cell walls and vacuoles, elements of the cellular osmoregulatory system, and form an extended and specialised cytoplasm. Several types of tapetum occur in different plant species. The cells may either remain as an inner lining to the anther loculus, forming a *secretory tapetum*, or they may migrate in between the developing microspores to form an *amoeboid tapetum*. Further, the amoeboid cells may fuse to form a multinucleate *syncytium*. All types are secretory and their activities are necessary for development of the microspores, since mutants with absent or defective tapetum fail to complete microspore development.

Some of the roles of the tapetum are known. Apart from its general nutritive role, it synthesises and secretes $\beta(1\rightarrow3)$ glucanase ("callase") into the anther loculus, where it dissolves the callose wall of microspores after meiosis (Plate 54a). It makes precursors for sporopollenin, which becomes deposited not only in the exine (see Plate 56), but also in orbicules on the locular face of the tapetum. It deposits proteins into the exine. Some of these are allergenic, to the discomfort of hayfever sufferers, others are hydrolytic enzymes, including esterase, phosphatase, ribonuclease, amylase. Some of these proteins may function in pollen-pistil interactions, including a type of compatibility mechanism that is mediated by factors from the sporophytic parent. The tapetum degenerates as the pollen approaches maturity and leaves several kinds of deposit on the grains. One consists of flavonoid pigments, another a mixture of carotenoids and lipids. These too may have roles in signalling between the pollen and the stigma during pollination. A third category looks like cellular debris and is known as *tryphine*. It is apparently crucial for hydration of the grains when they reach the stigma. Techniques for the localisation of gene transcripts and protein products will doubtless reveal further functions for the tapetum. In its active state, as shown here, it is quite unlike any other type of plant tissue.

Plate 55a-c This plate illustrates the cytoplasm of mature secretory tapetum in a species of oat, *Avena strigosa*. In several respects the cells resemble a polarised animal epithelium, with one face (on the right in (a)) directed towards the anther locule (LO), and the other (on the left) based on the cells comprising the anther wall (see also Plate 54). A small portion of a pollen grain intrudes at upper right. The specialised nature of tapetal cells is reflected in their markedly heterochromatic nuclei (Plate 54c), suggesting that many genes have been inactivated while the remaining euchromatin initiates a high level of activity for the dedicated tapetal functions.

Orbicules (O) and remnants of cell wall material (arrows) are present on the locular side of the tapetal plasma membrane (PM). Whole orbicules can be seen in Plate 56. They are initiated as spherical droplets of lipoidal material lying within cup-shaped depressions over the locule-face of the cell. These central cores become covered with sporopollenin with a structure similar to that of the tectum of the pollen grain exine. Their function is uncertain, but may be something to do with dispersal of mature pollen from the anther.

The tapetal cytoplasm is filled with plastids (P), mitochondria (M), Golgi bodies (G), and both rough and smooth endoplasmic reticulum (ER). The rough endoplasmic reticulum cisternae tend to lie in aggregates of fairly parallel cisternae. The cytoplasmic surfaces of the cisternae are covered in polyribosomes with their ribosomes arranged in spirals. The rough cisternae are continuous with smooth regions, as shown at the dashed lines in (a) and (b). Both types of cisternae contain a material of moderate electron density, presumably the products of the polyribosomes, but the smooth parts tend to be dilated into sausage-shapes or spheres (55c, arrowhead). Further narrow smooth cisternae ensheath plastids and mitochondria and link with the rough endoplasmic reticulum (best seen in 55b - arrowheads). These continuities are one of the best examples of regional differentiation within the endoplasmic reticulum system of a single cell. Given the large amount of lipid synthesis in the tapetum, it may be speculated that the cisternae that associate with plastids are receiving fatty acids for incorporation into complex lipids (see Plate 8).

The cytoplasm contains numerous large vesicles (V) with flocculent contents. Other micrographs suggest that these are derived from the Golgi apparatus, and profiles such as the one marked by an asterisk (lower left) that they exocytose their contents at the plasma membrane (not necessarily at any one face of the cell). There is no vacuole and since the cell walls have largely degenerated, the tapetum cannot be turgid. Its cytoplasm must be in osmotic equilibrium (isotonic) with the locular fluid.

The cell wall structure seen on the left hand side of 55a is complex. The plasma membrane (PM) is not in close contact with the wall, which probably consists of an electron-transparent lipoidal deposit (L) on an otherwise conventional type of wall (open arrow). Some plasmodesmata (PD) pierce both layers. The composite wall may create a sealed apoplastic compartment within which the tapetum can nourish the microspores without loss of nutrients that might otherwise leak from the locule. (a) x28,000; (b) x48,000; (c) x48,000.

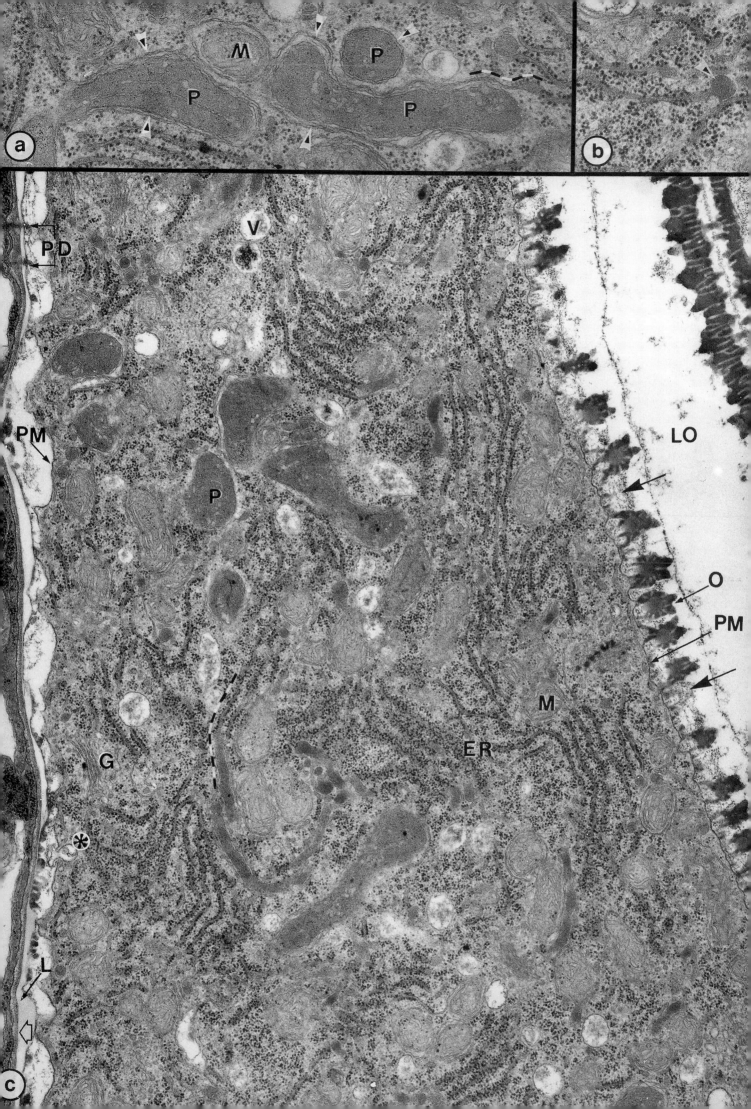

56   Pollen Grains (2): The Mature Wall

Introduction: The surface of the pollen grain wall is spectacularly sculptured. Patterns of exine deposition can be so species-specific that they can be used to identify pollen grains excavated from archaeological or geological remains. The patterns are well displayed by scanning electron microscopy, as in this plate.

Lilium longiflorum pollen, shown here, differs from the oat pollen in Plate 54c in lacking a continuous tectum (roof) on the exine. Also, whereas Plate 54 exemplifies those types of pollen that possess a pore or pores (*porate* grains), the lily pollen exemplifies a second main type, those possessing a less-ornamented, flattened region known as a *colpus* (*colpate* grains). The open wells created on the surface of grains with an absent or incomplete tectum provide reservoirs for the adherence of sporophytic factors from the tapetum. Lipidic and proteinaceous material helps pollen grains stick to the surfaces of insect vectors, and they may also contain recognition factors governing the success of the grain in gaining acceptance by the pollinated flower (Plate 58).

Plate 56a Pollen grains are seen lying in an anther loculus at the time of dehiscence. The remains of the tapetum and the loculus wall lie along the bottom of the picture. The surface of each grain is divided into a smooth region, the colpus, and a more extensive, elaborately sculptured region. The colpus is a thin area of exine on each grain and several examples showing its inward collapse are included. There is also a very small and shrivelled sterile grain. The colpus seems to be the spatial and functional equivalent of the pore seen in oat pollen in Plate 54b; it provides an exit point for the pollen tube during germination, and its buckling relates to pollen hydration and dehydration. In different species there may be one, three or six of these colpi per pollen grain. There is no surmounting tectum on the colpus. It is remarkable that whilst the position of the colpus is determined in relation to the cell axes in the tetrad, the genes that govern the detailed pattern of bacula and tectum exert their influence earlier, in the pollen mother cell, that is, at a stage similar to that in Plate 54a, prior to the appearance of either intine or exine. Once the genes have been transcribed, the information they contain is present in the cytoplasm, ready to be utilized at the appropriate time during pollen maturation. Thus experimentally produced fragments of cytoplasm from mother cells or young pollen will develop the patterned exine even if nuclei are absent. x640.

Plate 56b and c (b) shows part of a pollen grain (upper half of micrograph) lying on a carpet of orbicules (more or less spherical bodies, coated with sporopollenin, seen in the lower half of the micrograph). The layer of orbicules may aid the dispersal of the pollen when the anther loculus matures and splits open. In (c) the architecture of the baculae standing on the nexine can be clearly seen. The major difference between lily and oat

(Plate 54c) exines is that, in the former, the columnar bacula are not crowded all over the surface, but are restricted to the sides of irregular polygons. The tectum or roofing layer, on top of the baculae, of necessity lies in the same polygonal pattern, creating a balustrade-effect. Formation of the polygonal patterns of baculae seems to be governed, in some way, by the cytoskeleton and endoplasmic reticulum cisternae within the microspore. Some of the cisternae form a patchwork pattern just below the plasma membrane which precedes the polygons on the surface. The surface pattern of baculae is disrupted if anti-microtubule drugs are applied at early stages of development, but later it becomes fixed and insensitive to such treatments. (b) x4,500; (c) x13,500.

The great variety found in the morphology of pollen grains is based on the features outlined in this plate and Plate 54. The possesssion of pores or colpi, and the architecture of the tectum and its channels, all serve to create a wealth of structural variation. These structural features of the exine have been used to devise keys for the identification of plant species from their pollen grains. The resistant exines ensure that these features are preserved in grains that have fallen into lake sediments and peatland deposits. Cores from these locations can be removed, dated and analysed to determine the vegetation history of a region that can, in favourable circumstances, extend back for tens of thousands of years.

Sporopollenin is a class of wall matrix polymer rather than a compound of singular composition. It appeared early in the evolution of the plant Kingdom, and in extant species it extends from the algae and fungi through bryophytes and lower vascular plants to the angiosperms. A form of it is present in the walls of some unicellular algae (e.g. *Chlorella*, shown in Plate 18d), but usually it is limited to reproductive spores, where it protects against attack and decay while the spore passes through its hazardous dispersal and dormancy periods. Formerly it was thought to be formed by oxidative polymerisation of carotenoids and their esters, but more recent chemical investigations indicate that most sporopollenins may be built from long chain fatty acid precursors rather than carotenoids, though some samples do contain the latter. Thus sporopollenin is more similar than was realised to cutin and suberin, the other main protective compounds of plant surfaces (Plate 52). Subcellular details of its manufacture and secretion remain obscure, but there are general observations that its deposition is initiated upon thin lamellae which appear outside the plasma membrane at sites that define the future pattern and act as templates.

Micrographs kindly provided by J. Heslop-Harrison and H. Dickinson, reproduced by permission from: (56a) *27th Symposium Soc. Dev. Biol. Suppl.* **2**, 1968, Academic Press; (56b) *Planta*, **84**, 199-214, 1969; (56c) *Society for Experimental Biology Symposium*, **25**, 1971.

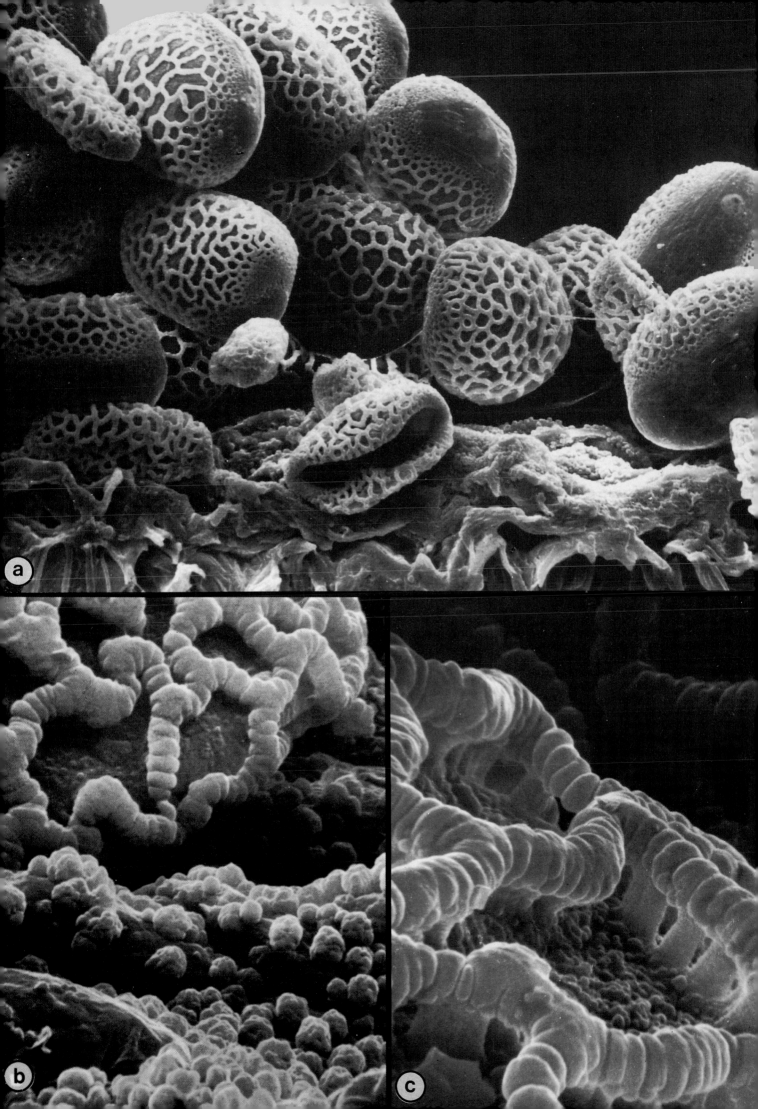

57 Pollination and the Generative Cell

Introduction: Pollination initiates specific interactions between pollen grains and the stigmatic surface and between pollen tubes of germinating grains and the tissues of the style. The stigma may receive bacterial and fungal spores as well as pollen from many plant species. Pollen of the correct, compatible, type is allowed to germinate and form pollen tubes that penetrate the stylar tissue. The stigma has broad-spectrum defence systems to prevent access of pathogenic agents to the reproductive structures of the ovary, based on toxic components such as protease inhibitors. The stigma and the transmitting tissue of the style, through which the pollen tube grows, also excludes pollen tubes from non-compatible pollen of the same or other species. The choice between nurturing of compatible pollen and inhibition of incompatible pollen depend on interactions between *recognition molecules* that express the genetic specificity of the stigma and style, and counterpart molecules carried by the pollen and present in the grain and/or the pollen tube. In some plants the latter arise from gene expression in the pollen grains themselves (gametophytic incompatiblity mechanisms) and in others from gene expression in the tapetum of the male parent (sporophytic incompatibility mechanisms). In sporophytic incompatibility the recognition molecules are released from exine cavities when the pollen grain lands on the stigma, so the compatibility-incompatibility decision is made very early, and either prevents or favours hydration and germination of the grain. In gametophytic control systems, incompatible tubes may grow into the style, but slowly, and are soon stopped by the stylar recognition molecules (probably ribonuclease enzymes with specific targets in the pollen tube cytoplasm).

Microspores undergo at least one round of mitosis, to give generative and vegetative cells, before being shed from the anther as pollen grains. It is the vegetative cell that germinates to form the pollen tube. In some plants mitosis of the generative cell, which lies within the vegetative cell, occurs before release of the grain (tricellular pollen), in others (bicellular pollen) it is postponed until after pollination and germination, when the generative cell has entered the growing pollen tube. The generative cell consists mostly of a nucleus, with little cytoplasm, which may be lacking in one or both of the heritable components of plant cells, mitochondria and plastids. If they are lacking, they are inherited maternally, *via* the population present in the egg cell. Mitosis generates two sperm cells lying close together in the original cavity within the vegetative cell. The two sperm cells may not be identical, one having more mitochondria or plastids than the other. In some plants they remain associated with one another, with intertwined cell extensions that also link to the vegetative cell, forming a complex known as the *male germ unit*. Sperm cells have a prominent cytoskeleton of microtubules, which defines the shape of each cell by forming a spindle-shaped basket around the nucleus. Generative and sperm cells have to be transported down the narrow pollen tube through the cytoplasm of the extended vegetative cell. Their cell walls, if any, are too tenuous and deformable to determine cell shape, so the arrays of microtubules have a skeletal role equivalent to the highly elaborate microtubule systems found in the cell body of wall-less sperm cells of lower plants (Plate 32a). The cells round up if their microtubules are removed.

Plate 57a Pollen grains adhering to stigma papillae of *Arabidopsis thaliana,* showing the patterned exine surface on each grain and the slits or colpi through which germination occurs. This species has three such slits on each grain (tricolpate). The grains adhere to the turgid papillae *via* their surface deposit of tryphine. When they germinate, the pollen tubes grow into and along the walls of the papillae (arrow). x120.

Plate 57b An example of self-pollination in a self-incompatible tomato plant. The stigma/style was squashed and stained with aniline blue, which renders the callose of the pollen tube walls and the sieve elements of the style fluorescent, as shown in this montage of 12 light micrographs. The style has two vascular bundles (large arrows). The pollen tubes grow through a tract of special nutritive cells, the *transmitting tissue*. Self-incompatibility barriers can be overcome in a number of ways, e.g. by pollinating an immature flower, which can allow the tubes to grow through the style before the female recognition molecules have been synthesised by the transmitting tissue. In this case a detached flower was incubated in the dark in distilled water and examined 48 hours after pollination. Most pollen tubes arrested in the top half of the style (open arrowhead), but some succeeded in growing about 8mm to the bottom (closed arrowhead). In the normal incompatibility response, all incompatible tubes stop at the top of the style. Kindly provided by M. Webb and E. Williams, reproduced by permission from *Ann. Bot.* **61**, 395-404, 1988. x19.

Plate 57c,d The generative cell of tobacco (*Nicotiana alata*) is moved into the pollen tube before it divides. These two micrographs show regions of the cell lying in the pollen tube (vegetative cell) cytoplasm. Fast freezing and freeze substitution have preserved the plasma membranes (PM) of each cell and the intervening thin cell wall much more faithfully than is achieved by chemical fixation of these delicate cells. The haploid cell nucleus (N), with pores (P) in the nuclear envelope, occupies most of the cell volume. The cytoplasm is dominated by an extensive microtubule cytoskeleton (MT), seen in transverse section in (c) and longitudinal section in (d). Part of the vegetative cell nucleus (VN) is included, lying close to the generative cell in a male germ unit. Kindly provided by S. Lancelle and P. Hepler. (c) x47,500; (d) x42,500.

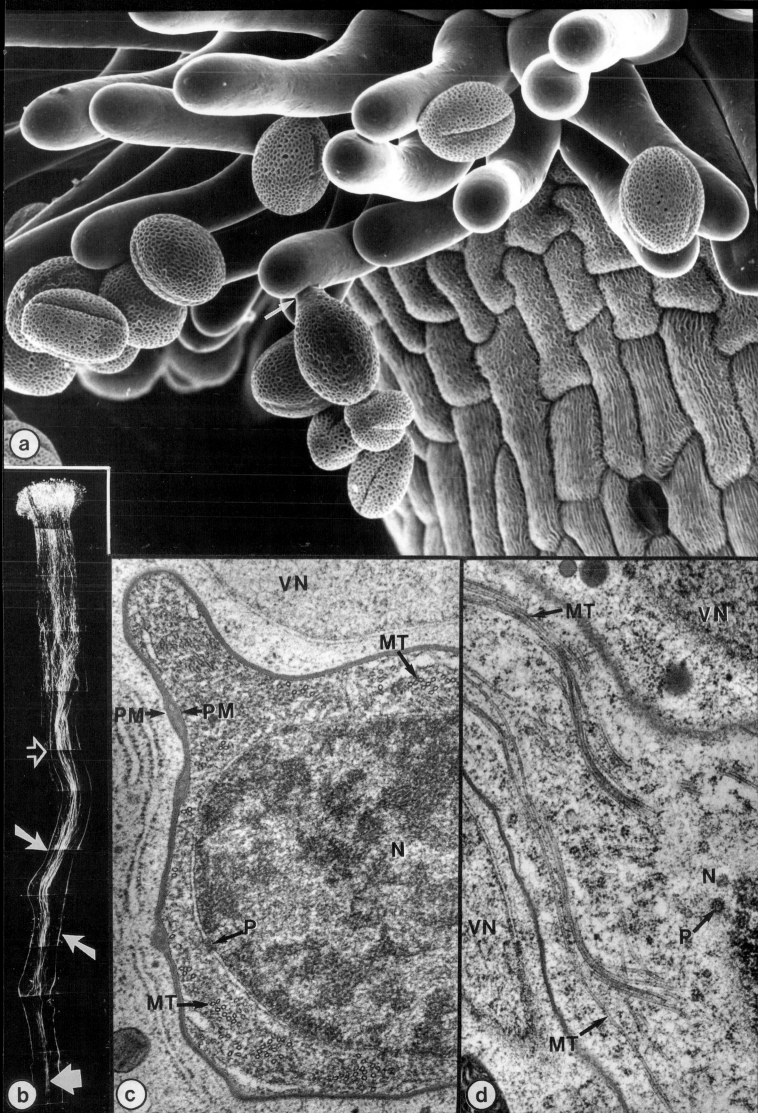

58 Pollen Tube Cytoplasm

Introduction: Pollen tubes are vital components of the reproductive process in flowering plants. Their role is to deliver sperm cells to the embryo sacs of ovules, following delivery of pollen grains to the stigma by agents of pollination (wind, insects, etc.). They grow into the stigmatic tissue and then between cells of the stylar transmitting tissue to the ovary. Depending on the form of the ovary, they then exit from the placenta and cross an air gap to reach the micropyles of ovules. They enter the embryo sac through a synergid cell. The tip of the pollen tube ruptures and the sperm cells are released (Plate 59).

The whole chain of events leading to the fertilisation process is highly competitive, and the severe selection pressures to which pollen tubes are subjected throughout their growth filter out many of the less-fit genotypes that were produced by segregation of genes in microspore meiosis. Incompatibility mechanisms (see Plate 57) filter out pollen tubes which, if successful, would lead to inbreeding. Consider maize: pollen that survives desiccation during airborne transmission to a stigma germinates within minutes of landing and uses previously stockpiled reserves of cell wall precursors, stored in vesicles, to grow about 30cm in 24 hours. It is hard to find examples of comparable growth rates anywhere in the plant or animal kingdoms. In general there is not enough cytoplasm in a vegetative cell in a pollen grain to fill the volume of a fully-grown pollen tube, and this deficiency is circumvented by a unique mode of growth. Unlike most plant cells, pollen tubes grow at their tips, extending at rates up to several micrometres per minute. Their cytoplasm shows very rapid streaming, and stays near the advancing tip. The tube that is left behind is sealed off at intervals by deposition of plugs of callose (which is also the major constituent of the tube wall, and when stained with aniline blue, reveals the length and position of the tubes in the style, as in Plate 57b).

Pollen tubes are easily damaged by the conventional procedures of specimen preparation for electron microscopy. Such damage was minimized in the examples shown here by freezing un-fixed pollen tubes very rapidly in liquid propane cooled to the temperature of liquid nitrogen, followed by freeze-substitution. Other examples of this superior method are shown in Plates 13a, 46a and 57c and d.

Plate 58a This low magnification view shows part of a pollen tube in longitudinal section. In life, pollen tube cytoplasm streams very rapidly and even with the best freezing rates some of the components may move slightly before they are locked in place by vitreous ice formation. The longitudinal stratification of components, seen in the micrograph, conveys the impression of a flow of cytoplasm along the mid region and slower eddy currents nearer the sides of the tube. Such flows are indeed seen in living tubes. The mid region shows numerous cisternae of endoplasmic reticulum, nearly all oriented in the (longitudinal) direction of cytoplasmic flow. So too are the more densely stained mitochondria. The bulkier plastids, mostly containing starch grains, and numerous small vacuoles, are more peripheral, though dense cytoplasmic ground substance extends between them out to the cell wall. Golgi stacks are also abundant. Vesicles are often found in chains, also lying in the axis of the main flow. *Nicotiana alata*, x5,100.

Plate 58b The rapid streaming of pollen tube cytoplasm is driven by an actin-myosin system (see also Plates 31 and 35). This *Petunia* pollen tube was dehydrated and embedded after fast freezing to preserve antigenicity of its proteins. The micrograph shows images of microtubules (MT), mitochondria (M) and ribosomes in the background cytoplasm. Staining and contrast are impaired, but the presence of actin can be demonstrated by labelling sections with a rabbit anti-actin antibody which is in turn detected with goat anti-rabbit antibody labelled with 10nm gold particles. Gold label (arrows) occurs over the cytoplasm, including alongside one microtubule, a location where actin filaments are frequently found (Plate 31a). The antibody labels only those actin molecules that lie with their antigenic sites exposed at the surface of the section. Kindly provided by F. Doris, x80,000.

Plate 58c This high magnification detail of the same material as in (a) is representative of the motility apparatus. The main feature is a bundle of microfilaments of actin. In this example, endoplasmic reticulum, vesicles and a mitochondrion lie in contact with the actin bundle. Careful inspection shows small, fuzzy spokes between the actin bundle and these associated structures. These connectors may well be myosin (and associated) molecules on the surface membranes of the streaming objects. Energy in the form of ATP is used to drive conformational changes in the actin-myosin interaction, such that the objects are propelled along the actin bundle (see also Plate 35). *Nicotiana alata*, x55,000.

Plate 58d A mass of vesicles and membranes accumulates at the growing tip of pollen tubes. In this region of very dynamic activity in the pollen tube, vesicles fuse with the plasma membrane and liberate their contents to augment the cell wall (inset). There is a local, high concentration of calcium ions in the cytoplasm, associated with vesicle fusion and tip growth. If the calcium gradient dissipates, the vesicles disperse and growth stops. The cell wall at the growing tip is easily deformed or burst. Unlike other plant cells, which make callose when wounded, pollen tubes have active callose synthetase as part of their normal complement of enzymes. The pollen tube wall is strengthened by deposition of callose, starting just behind the growing tip. *Lilium longiflorum*, x10,500, inset x47,000.

(a), (c) and (d) kindly provided by S. Lancelle and P. Hepler.

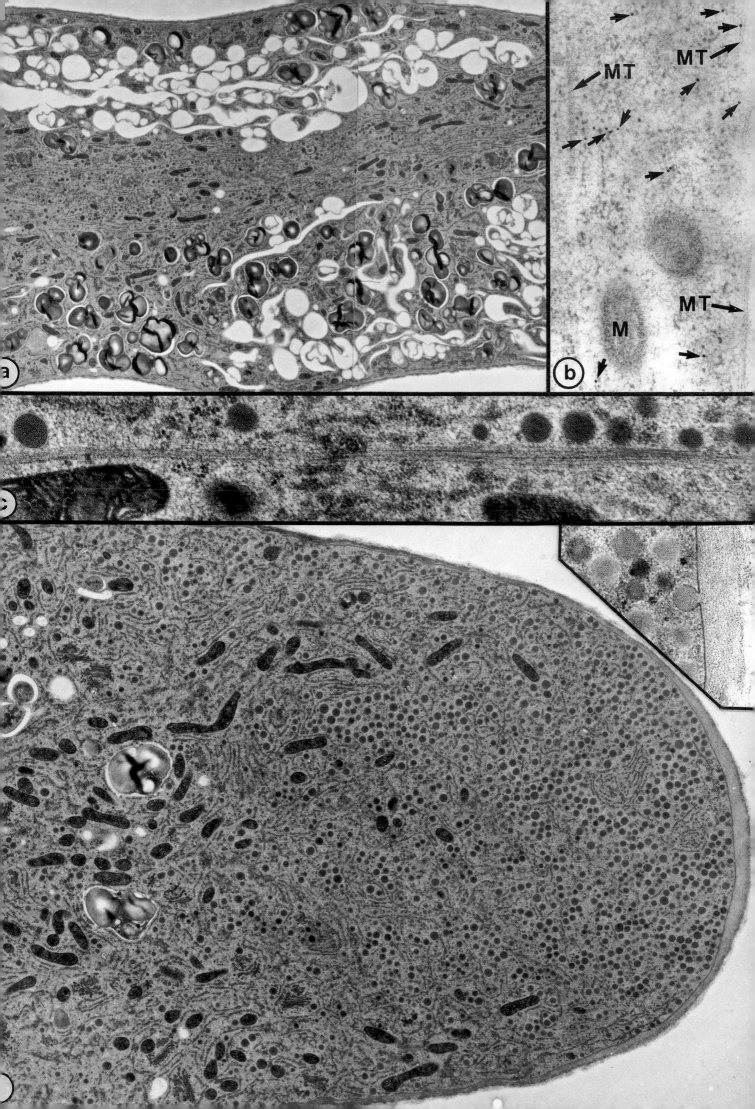

59 Female Reproductive Tissues and Embryogenesis

Introduction: In female reproduction, *megaspore mother cells* undergo meiosis to give rise to four haploid cells. Only one survives as a functional megaspore. Whereas many pollen grains develop in each anther (Plates 54, 56), megaspores develop singly in ovules, of which there may be many in the *ovary*. Each megaspore produces an 8-nucleate (usually) embryo sac, the mature female gametophyte. One of the 8 nuclei is in the egg cell. Pollination delivers pollen grains to the stigma and pollen tube growth down the style (Plate 58) delivers two sperm cells to an embryo sac, where during fertilisation one sperm nucleus fuses with the egg nucleus and the other sperm nucleus fuses with two *central cell* nuclei. The fertilised egg, now the diploid *zygote*, develops into the embryo, and the triploid product of the central cells and sperm divides successively to generate nutritive endosperm tissue. In most cases, seeds are made up of embryo, endosperm and protective tissues derived from the maternal *integument* tissues of the ovule wall. Seeds develop in a fruit, the product of the ovary.

Plate 59a: This electron micrograph shows a longitudinal section of a young ovule of the orchid *Stenoglottis,* containing a large megaspore mother cell in prophase of meiosis. The jacket of cells around the megaspore mother cell is the *nucellus*. Arrows show the *inner integument,* which will grow opposite the direction of the arrows to ensheath the embryo sac, leaving only a narrow pore, the *micropyle*, for entry of the pollen tube at fertilisation. x3,000, kindly provided by S. Tiwari.

The remaining pictures are of *Arabidopsis thaliana.* (b,c,e,f,g) kindly provided by M. Webb.

Plate 59b,c: Scanning electron micrographs of early ovule development. (b) is at the same stage as (a) and shows the nucellus, containing the megaspore mother cell, protruding from the young *inner and outer integuments*. x200. (c) is a later stage showing how the integuments have grown over the nucellus. They ensheath the megaspore (and the future embryo sac), leaving an apical pore, the *micropyle*. The ovule stalk often becomes curved, and in *Arabidopsis* it bends through 180° so that the apical (micropylar) end of the ovule eventually faces the base of the ovule stalk. x100.

Plate 59d: The inset (lower left) shows a post-meiotic stage in a whole, cleared ovule, x600. One of the four products of meiosis becomes the functional megaspore (arrow) and the other three (arrowheads) degenerate. The other pictures show a cleared ovule. The megaspore has undergone successive mitoses to produce the embryo sac. The inset at upper right shows the recurved ovule (x280) and the main picture (x2,500) the micropylar end of the embryo sac (M - micropylar pore). The most conspicuous features of the embryo sac nuclei are their nucleoli. Two *synergid cells* lie at the micropylar end; in this view just one nucleolus is visible (S). The synergids receive the incoming pollen tube. Although there is no well-formed

cell wall, the outline of the egg cell is visible (arrows), as is its prominent nucleolus (E). The other two nucleoli (C) belong to the two central cell nuclei, which lie in a common cytoplasm. They have been brought into the same apparent focal plane here by video superimposition of two different levels. The refractile granules are starch grains, already accumulating in the central cell, which is the progenitor of the endosperm tissue. *Antipodal cells* lie at the other end of the embryo sac, out of this view.

Plate 59e,f,g: The embryo sac enlarges after fertilisation. (e) shows a section that includes an elongated but still single celled zygote (Z) in a large vacuole. (f) and (g) show early stages of endosperm formation, with telophase stages of division (arrows) in (f) generating wall-less amoeboid cells in (g) (see Plate 38). (e) x350, (f,g) x670.

Plate 59h-p: This developmental sequence shows micrographs of embryos within cleared ovules (or, in (n-p), embryos that have been isolated from ovules). Preprophase bands of microtubules have not featured in meiosis or embryo sac divisions, but they reappear at the first division of the zygote, the first where a new wall is laid down in a predetermined plane. After elongation (see (e)) the zygote divides transversely near its apical end. The densely cytoplasmic head becomes the *pro-embryo* and the stalk the *suspensor*. The suspensor divides transversely several times (h-m). The proembryo undergoes a regular sequence of divisions. The first is longitudinal in one plane (i), then longitudinal again, but in the plane at right angles to the first division, then transverse again to make an *octant* stage (j). A set of tangential divisions then generates the future epidermal cell layer (k). Divisions in the interior and divisions in the surface layer perpendicular to the surface (l,m) produce the *globular embryo* stage. Longitudinal and transverse divisions in the centre of the globular embryo just above the attachment of the suspensor now prepare the future root apical meristem. The topmost cell of the suspensor is called the *hypophysis*. As shown at higher magnification in (n), it divides transversely to make two cells whose derivatives are important in organisation and growth of the root meristem. Meanwhile divisions concentrate in two flank regions to initiate the two cotyledons in the early *heart* stage embryo (o). The main axis of the embryo is now consolidated. The late heart stage (p) has a distinct root meristem and two cotyledon initials, between which lies the future shoot apical meristem. (h,i) x750; (j-m) x500; (n,o) x1,000; p x350.

Plate 59q: A cleared root apical meristem of an embryo from a seed shows how the embryonic cell patterns establish future tissue organisation. The central root cap is derived from the lower derivative of the hypophysis (bottom of (o)). Lateral derivatives of the upper part of the hypophysis propagate the cortex and endodermis. Longitudinally-elongated cells, first seen in (n), are progenitors of the root stele tissue. x1,000.

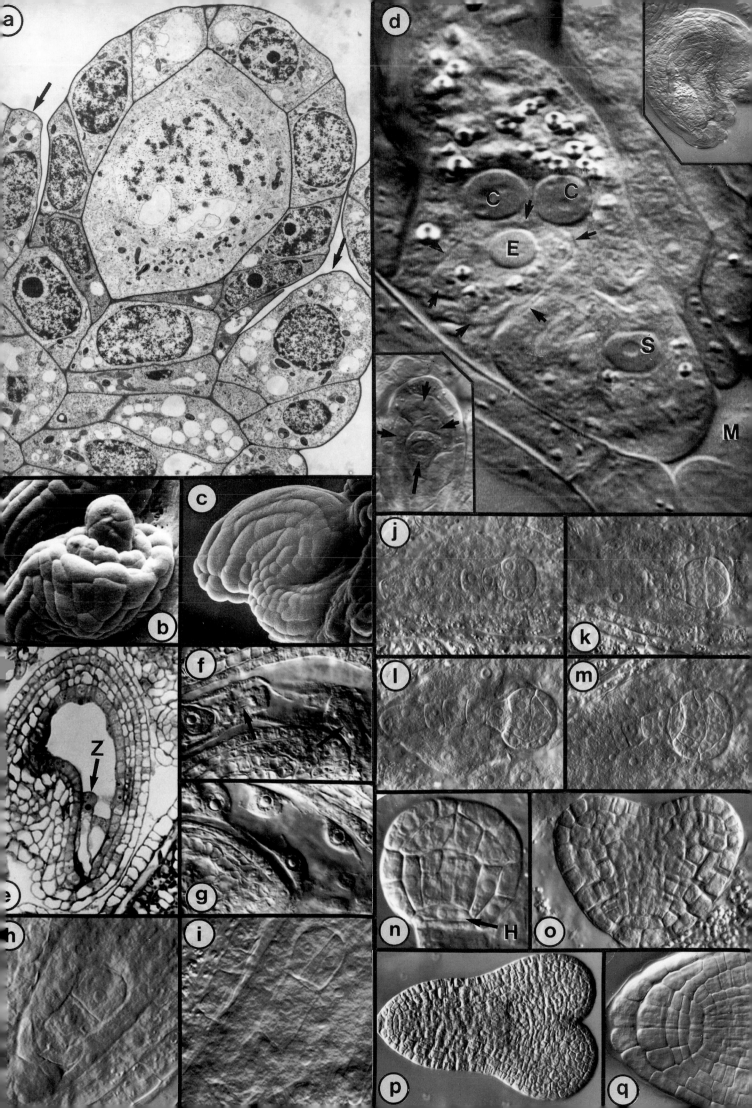

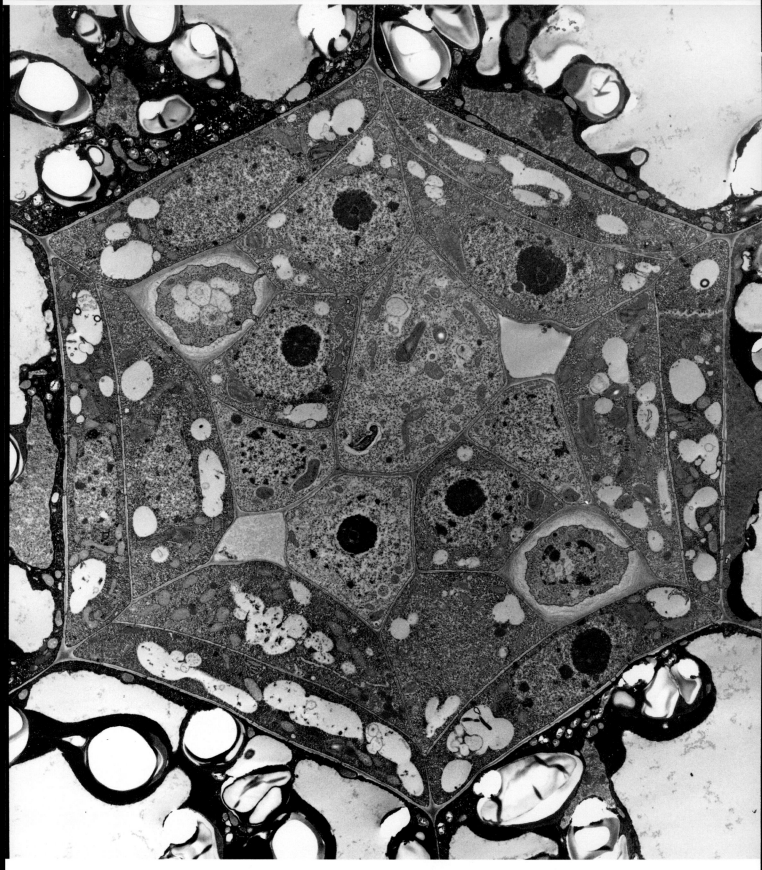

Plate 60: Plates 1-59 have focused on cells and cell components. Plate 60 highlights a higher level of organisation in plant structure-function relationships.

The cells seen here are juvenile, in a cross section of the zone of cell differentiation near the apical meristem of the unusually miniaturised root of the water fern *Azolla* (x5,800). Despite their close proximity, the 22 cells of the differentiating stele are proceeding along six different pathways of maturation in a precisely defined pattern, producing endodermis, protoxylem, metaxylem, sieve elements, phloem parenchyma, and pericycle. The mature state is shown, in the same orientation and with cell types labelled, in Plate 51a. The six cell types collaborate to perform the many functions of the root. Each plays its distinctive part, but no one could survive on its own. Just as specialised *sub*-cellular components participate in a collaborative group existence upon which depends the survival of the whole cell, so it is with cells in plant tissues, organs and the whole plant. *Supra*-cellular collaboration creats functional entities at levels of organisation higher than that of the individual cell - and beyond the scope of the present book.

Index

Numbers refer to Plate numbers, those prefixed with the letter 'i' refer to page numbers in the Introduction.

Plant Cell Biology

BRIAN E. S. GUNNING / MARTIN W. STEER

STRUCTURE AND FUNCTION

Tremendous advances have been made in techniques and application of microscopy since the authors' original publication of *Plant Cell Biology, An Ultrastructural Approach* in 1975. With this revision, the authors have added over 200 images exploiting modern techniques such as cryo-microscopy, immuno-gold localisations, immunofluorescence and confocal microscopy, and in situ hybridisation. Additionally, there is a concise, readable outline of these techniques.

With these advances in microscopy and parallel advances in molecular biology, more and more exciting new information on structure-function relationships in plant cells has become available. This revision presents new images and provides a modern view of plant cell biology in a completely rewritten text that emphasises underlying principles. It introduces broad concepts and uses carefully selected representative micrographs to illustrate fundamental information on structures and processes.

Both students and researchers will find this a valuable resource for exploring plant cell and molecular biology.

Special Features

- Content of illustrations and micrographs expanded to over 400
- Incorporates 20 years of technological advances in microscopy
- Combines pictorial and textual strengths previously found in separate volumes
- New! Four pages of color plates, highlighting confocal microscopy

About the Authors

Brian E. S. Gunning has a distinguished record of research publications in the field of plant cell biology, especially using approaches based on microscopy. His career began in Queens University, Belfast; since 1974 he has been a Professor in the Institute of Advanced Studies of the Australian National University. His work has been recognised by his election to national scientific academies in the UK and Australia, and by awards from the American Botanical Society and the Linnaean Society.

Martin W. Steer is Professor and Head of the Department of Botany, and Director of the Electron Microscopy Laboratory, University College Dublin. A graduate of Bristol University, he received his Ph.D. from Queen's University, Belfast, where he remained on the faculty for the first half of his career. He is an elected Member of the Royal Irish Academy, and holds editorial positions in this and other institutions. Formerly President of the Microscopical Society of Ireland, he is presently its Secretary.

Jones and Bartlett Publishers
40 Tall Pine Drive
Sudbury, MA 01776
800 832 0034 • 508 443 5000

ISBN: 0-86720-504-0